Michael C. Thomsett
Math for Managers

Michael C. Thomsett

Math for Managers

—

Second Edition

DE|G PRESS

ISBN 978-1-5474-1670-7
e-ISBN (PDF) 978-1-5474-0063-8
e-ISBN (EPUB) 978-1-5474-0065-2

Library of Congress Control Number: 2018955034

Bibliographic information published by the Deutsche Nationalbibliothek
The Deutsche Nationalbibliothek lists this publication in the Deutsche Nationalbibliografie;
detailed bibliographic data are available on the Internet at http://dnb.dnb.de.

© 2019 Michael C. Thomsett
Published by Walter de Gruyter Inc., Boston/Berlin
Printing and binding: CPI books GmbH, Leck
Typesetting: MacPS, LLC, Carmel
www.degruyter.com

About De|G PRESS

Five Stars as a Rule

De|G PRESS, the startup born out of one of the world's most venerable publishers, De Gruyter, promises to bring you an unbiased, valuable, and meticulously edited work on important topics in the fields of business, information technology, computing, engineering, and mathematics. By selecting the finest authors to present, without bias, information necessary for their chosen topic *for professionals*, in the depth you would hope for, we wish to satisfy your needs and earn our five-star ranking.

In keeping with these principles, the books you read from De|G PRESS will be practical, efficient and, if we have done our job right, yield many returns on their price.

We invite businesses to order our books in bulk in print or electronic form as a best solution to meeting the learning needs of your organization, or parts of your organization, in a most cost-effective manner.

There is no better way to learn about a subject in depth than from a book that is efficient, clear, well organized, and information rich. A great book can provide life-changing knowledge. We hope that with De|G PRESS books you will find that to be the case.

DOI 10.1515/9781547400638-002

Contents

Introduction: The Basic Problem with Numbers

In this second edition of *Math for Managers*, an effort has been made to maintain a logical sequence of chapters and topics; and to keep all math on a basic level.

Most managers do not hold advanced mathematical degrees, so simplicity is a worthy goal. Many math books are overly complex and can be appreciated only by a limited number of managers. The author is not a mathematically inclined person and has not been educated in this area. However, he has considerable experience in management. Like most managers, he has had to struggle to master essential mathematical principles and guidelines, and this has not been the easiest attribute of his experiences. It is certain that the same holds true for most other managers, educated in business planning and marketing, but not necessarily in math.

Even so, the realities of the "mathematical world" cannot be overlooked. Business operates on measurement, and that invariably involves some level of calculation. *"It's all in the numbers."* Everyone has heard this statement and it is true. Your performance is invariably judged by how much profit you create or by how much cost you incur in your segment, team or department. The "bottom line"—profit or loss—is the universal means for monitoring performance and for determining whether an initiative was worthwhile.

With the dominance of the bottom line in every aspect of how your performance is graded, you have a distinct advantage if you are skilled at conveying information in terms of profitability and other numerical outcomes (budgets, customers, units of production, and sales, employees, etc.).

You are also at a distinct disadvantage if you cannot communicate the profit or loss aspects of your work to management. On a most basic level, it is less effective to just ask management for something, than to demonstrate how an approval is going to create additional profits or cut costs and expenses. This is the rudimentary distinction between managers with communication skills, and those who struggle every day trying to find the best way to communicate what they know.

If you do not have a background and an education in finance, you probably struggle with these issues daily. Even those trained in accounting may find it difficult to summarize their requests in plain, simple, and clear terms for management. No one is immune from the difficulty in matching numerical information with a concise request or recommendation. For some, the numerical aspects of the job are comfortable but conveying the essence to management is much more difficult. For others, even those with exceptional communications skills, reducing the numbers ("crunching") to the basics is the real challenge.

Your purpose in making effective use of numerical information is to convey essential data management needs to make an informed decision, and to make your case to management convincingly. Management faces an unending array of choices, and the desire is to make a choice that is not only most profitable, but that also involves less risk than other choices.

DOI 10.1515/9781547400638-004

It is not enough to demonstrate that a decision is likely to be profitable if it also incurs unacceptable risks. These include potential liability, supply chain losses, reduced customer satisfaction, or damage to brand and reputation for the company. When an esteemed company like Mattel contracted manufacturing in China, but failed to properly supervise quality control, the consequence was that toys were sold in the U.S. containing harmful lead. The product cost aspects of this mistake were easily rectified. However, the reputation to the company, while less tangible, is likely to affect profits at an unknown level and for an unknown period. The analysis of risk involves both tangible and intangible considerations, making it difficult to *know* how much risk is really involved in creating "*x* profits" as the result of "*y* decisions."

What this means for you is that any communication is going to be based on evaluation of profit and loss, many forms of risk, and the time required for return on investment, just to name a few considerations. How do you communicate the relevant facts to management? How do you reduce the research to well-supported recommendations or to caution statements? These are only a few of the issues you face in managing information and in massaging it to create an effective, simple, and honest method of communication.

This book is designed to interpret and simplify basic skills in using financial and statistical facts, whether in reports or presentations. The purpose in presenting information is not to provide instruction in communication abilities, but to demonstrate how a range of formulas are developed and how they apply to common situations. This will enable you to more confidently and effectively include financial and statistical data in your reports and presentations; and to provide convincing arguments to support your recommendations.

Each chapter introduces a series of formulas and explains the context in which they are applied. The step-by-step formulation is followed by an example to show how application of the formula works. Because so much of business involves management of money, the first two chapters deal with the time value of money, compound interest, and present and future value. Even though many of the calculations involved can be performed with handheld devices on online free calculators, you are going to be able to express these concepts more effectively by understanding how the calculations are performed, why time does create value, and how compounding works.

The next two chapters (three and four) involve calculating return and breakeven points, issues that virtually every manager (even those lacking financial training) must contend with. For many non-accounting managers, these topics can be overwhelming and complex, making effective communication a hopeless cause. However, the formulations themselves are not at all complicated; they are simply not explained well in most applications. These chapters show you how to make the complex simple and how to create powerful communication tools by mastering their explanations.

Chapters five through seven focus on financial reporting, known to be the realm of accountants. Why include these? For non-financial managers, the mysterious and complex methods of financial analysis take a lot of advantage away and make communication difficult. The skills needed to understand and explain financial reports are not really at all difficult; they simply need to be placed in context. This is especially true for topics such as depreciation and how this affects financial reports on a non-cash basis.

The next three chapters (eight, nine and ten) explore the preparation and organization of reports, budgets, and forecasts. Virtually every manager is required to prepare reports, create departmental budgets or forecast future revenues. Many dread this experience. These are not only numerically complex tasks but may also be political minefields within the organization. These chapters explain how to incorporate numerical information into reports, not as necessary evils that can destroy an otherwise effective presentation, but as the means to support strong arguments and *improve* how information is conveyed. The dreaded budget and forecast are explored with the same premise: Once you master the communication of the numbers, all your presentations (especially budgets and forecasts) become more effective and useful.

Chapter 11 demonstrates how to best use statistics, especially in reports. Most people dread statistics because they are complex. In this chapter, you will see how effective use of statistics can reduce pages of complex calculations down to a declarative sentence; and how to support conclusions with the numbers rather than obscuring the facts. The effective use of statistics creates insights among the audience for your reports, rather than the glazed-over eyes so often seen while presenters begin going through the numbers.

The last chapter (12) goes through a few effective and intriguing math shortcuts. These are useful in developing your mastery over the numbers and will also speed up the time required for mental calculation. Math may not always be an enjoyable exercise for those who struggle with it, but these shortcuts make the process more manageable and show how to convert a struggle into a series of rewarding exercises with satisfying results.

At the end of the book, all the formulas are summarized in alphabetical order for easy reference. This format compliments the organization of the book. Within each chapter, the discussion is always based on context, followed by a formula, and concluded with an example. The set of formulas, provided together in the appendix, helps to cross-reference the many formulas in one place.

The last section is a glossary of terms.

The purpose in preparing this book is to provide alternative methods for reports, budgets, and other numerical chores to make them effective and easier to communicate with others. The perception that a background in finance gives some managers an advantage over others is based on one reality: Training and experience pay off. This book closes that gap to a degree by arming you with communication skills designed to improve your reports and presentations. Once you master the numerical aspects of your message, your overall communication skills rise, and your effectiveness improves as a direct result.

Notations Used in this Book

An effort has been made to maintain a layman's version of mathematical functions. The standard symbols for basic functions are used in all formulas:

+   addition
−   subtraction
+(−) addition or subtraction
×   multiplication
÷   division

Symbols used for basic functions on spreadsheets conform to the Excel formatting, which uses the symbols, +, −, * and / for the four functions.

Superscripting is used to indicate multiples of a value. So n^2 is the direction for squaring n. In formulas requiring many more functions than squaring, the superscripted value is either shown or marked as an unknown number like n^x with x representing a variable. Subscripting is used to distinguish values from one another in a series of different values. When you see n_1 and n_2 you know that there are two different versions of n.

Parentheses are used in formulas where different functions must be separated. For example, the instruction $a + b \div c$ could be interpreted in two ways depending on how the operations are separated. Thus, $(a + b) \div c$ is not the same as $a + (b \div c)$. However, the rule here is that while you try to go from left to right, you have to multiply or divide first before you add or subtract. So, $a + (b \div c)$ is correct. If you try this on a calculator going from left to right, you will get the wrong answer.

Other mathematical expressions in use are the square root symbol, $\sqrt{}$ and two Greek symbols used in some mathematical functions, lower-case sigma, σ and pi, π.

In some instances, a symbol has been used to indicate affixing one value to another, using a colon. If you are to affix y to x, it is shown as $x{:}y$. In other words, rather than adding, digits of an affixed value are added to the end. For example, adding 25 to 700 creates the answer 725. Affixing 25 to 700 creates the result 70,025 and is shown in formulation as 700:25. There are other uses of the colon in mathematics and it is important to understand that in this book, the colon does not indicate a ratio.

The use of these shortcut symbols has been held to a minimum on the assumption that most managers prefer to avoid higher mathematical formulas or explanations that assume too much math training or knowledge.

Chapter 1
Compound Interest: The *Power* of Money

Money and time are directly and inescapably related. This means that the longer money is left on deposit, the more it earns (or, the longer it takes to repay a loan, the more it costs). Although this concept—that the benefit or cost of money increases over time—is easily explained but not always universally understood. This chapter explains how the time value of money works and provides formulas for calculating interest in various ways.

In the calculation of interest cost, time is the most critical element, even more so than the rate. These two factors—time and rate—define the true "cost of money." When an organization borrows money through working capital loans, equipment financing, or for any other reason, there is a tendency to focus on the interest rate only. While the rate is important, there is more to consider, including the monthly payment required and the length of time it is going to take to retire the loan. At 7.5%, for example, a 10-year repayment is going to cost twice as much in interest as a loan for the same amount with a four-year repayment.

Example: You borrow $20,000 from your local lender. You have a choice: Repayment in four years at $483.58 per month, or repayment in eight years at $277.68 per month. Your first reaction is that the lower payment is desirable. However, when you add up your total of payments for each of these loans, you discover the truth: The total for the four-year term is $23,211.84 (48 months x $483.58); and the total for the eight-year term is $26,657.28 (96 months x $277.68). The difference in total interest is $3,445.44. The interest cost for the longer-term loan is twice as much as for the shorter-term loan.

Selecting a repayment period is a matter of balance between the affordability of the monthly payment and the overall cost of interest. This is the essence of the time value of money. So, in calculating what a repayment will cost for this $20,000 loan, you need to evaluate the interest rate and monthly payment; however, you also need to compare the total cost of interest based on different loan repayment terms.

In addition to the monthly payment and overall interest cost, the method of interest calculation is going to affect the total of payments as well. You need to employ different compounding methods of interest for a variety of reporting and budgeting concerns, not to mention calculating the cost of borrowing money.

Time Value of Money—The Concept

The combination of elements defines the true cost or benefit of money. The time value of money "... represents the building block and basic tool for many other fundamental topics such as bond valuation, stock valuation, capital budgeting, and options

DOI 10.1515/9781547400638-001

valuation."[1] The cost of money is incurred when you borrow and the benefit results from savings. There are four of these elements:

1. *Amount borrowed.* The most easily understood element of all is the amount borrowed. Most people understand that the more money borrowed, the higher the repayment is going to be. This is obscured, however, by the varying payment levels for different lengths of repayment. For example, at 7.5%, $20,000 requires monthly payments of $483.58 over four years. However, you can borrow $30,000 and pay only $416.52 per month, or *less* in monthly payments. The drawback, however, is that it will take eight years to repay the $30,000. Total interest is $9,985.92 (almost the same difference between the four-year $20,000 loan and the eight-year $30,000 loan). The smaller loan with faster repayment costs $3,211.84, or interest of about one-third of the longer-term loan with smaller payments.

    Which loan is best suited to your needs? For most business owners and managers, the commitment to debt service twice the length of the original $20,000 has to be a primary consideration. The amount borrowed is $10,000 more, but you are committed to repayments for twice as many years. It is easy to develop rationale justifying this lengthier borrowing schedule. For example, if you originally wanted only $20,000, why not borrow $30,000 and invest the difference? The payments are about the same amount, but the $10,000 is enough to repay all the interest on the higher loan. This argument overlooks two important facts, however. First, although the higher loan amount creates enough cash to pay the interest, you also must repay the additional $10,000 borrowed, and that translates to twice the length of repayment. Second, will you really save the difference? As many business managers have realized, it is difficult to set up a reserve and leave it in place. Over time, the temptation to use that fund for other necessities is going to recur, and ultimately, the result is the same: The longer-term loan is going to be more expensive and require a lengthier repayment commitment.

2. *Repayment term.* The question of what repayment term to pick should never be based on the monthly payment alone; it should also include an analysis of cash flow requirements and limitations (see Chapter 4), and the affordability of borrowing. You may want to borrow money for any number of reasons, but all should be analyzed with a series of key questions:
    -   Can I afford the repayments?
    -   How does a loan affect my cash flow?
    -   Have I identified how the loan will increase profits? (This may occur via expanded markets, greater efficiencies, or improved products or services.)

---

1 Martinez, V. (2013). Time Value of Money Made Simple: A Graphic Teaching Method. *Journal of Financial Education, 39* (1/2), 96–117.

The repayment term might seem like a no-brainer in the sense that you want to get a loan repaid as quickly as you can afford, for the lowest interest cost and least impact on cash flow. However, the question also has to depend on affordability and cash flow, and not merely on the concept that "more is good" when it comes to adding debt. This common belief can be destructive not only to your ability to fund repayments while maintaining cash flow, but also on how much negative impact debt might have on future expansion and profits.

3. *Interest rate.* The interest rate you are required to pay to borrow money (or that you are paid to save or invest) makes a tremendous difference over time. Some loans can be negotiated for a lower interest rate in exchange for more rapid repayment, saving money over the full term. For example, the difference between 7.0% and 7.5% is about $5.19 per month over 10 years. For a $20,000 loan, that comes out to a difference of $622.80. For a $200,000 loan the difference is about $6,228 for that one-half of one percent difference in the rate. So, negotiating a rate downward by a half percentage point makes a difference; and the larger the loan, the more the dollar value of the savings.

   The interest rate can also be either fixed or adjustable. Although these terms are most often associated with residential mortgage loans, they can also be applied to business loans of many types, and with varying terms. An interest-only loan can be renegotiable after a few years; however, the rate you will be expected to pay is likely to change based on the interest market at the time. In this respect, the interest rate—unless fixed for the full term of the loan—is the great variable in this evaluation.

4. *Compounding method.* The previous cases have all been based on monthly compounding of interest. This means the "nominal" rate (the rate stated by the lender) is divided by 12 (months), and the resulting monthly interest is calculated against the current loan balance. This results in an annual rate higher than the nominal rate. The higher your interest rate, the more expensive monthly compounding is going to be.

   Compounding is not mysterious, but it often is expressed poorly so that many (including managers) do not have a working knowledge of this process:

   Practitioners in the finance, accounting, and legal professions are often required to use return or interest-rate quotes to determine interest charges, present value and future amounts. Unfortunately, ... interest rates are rarely quoted in a form than can be used directly in basic time-value calculations.[2]

Banks may charge monthly compounding rates for money they loan, while paying you only quarterly compounded interest for funds you leave on deposit. While this is not

---

2 Stangeland, D., & Mossman, C. (2006). An Effective Method for Teaching and Understanding Interest Rate Conversions. *Journal of Financial Education, 32*, 97–114.

equitable, the banks also know that *you* need the loan at least as much as they want to grant the loan. Most managers pay little attention to the compounding method because it does not make much difference in the actual rate. For example, 7.5% compounded monthly comes out to an annual rate of 7.76% (to calculate compounding, see the section later in this chapter). In comparison, quarterly compounding produces an annual rate of 7.71%, or only five one-hundredths of a percent less. The difference over the repayment term of a loan adds up.

Simple Interest

To calculate interest, whether on a loan or a savings account, the basic formula—simple interest—is easy. Simply multiply the stated interest rate by the principal amount (the amount borrowed). The application of simple interest is that it is calculated on principal only. This definition implies "that if a loan is repaid in instalments, equal or unequal, no part of any installment may be considered as an interest payment if simple interest is specified."[3]

This distinction defines the difference between simple and compound interest. To begin, the formula for simple interest involves only one step.

Formula: Simple Interest

$$P \times R = I$$

where: P = principal
R = interest rate
I = interest

On a spreadsheet, enter the following:
A1   P
B1   R
C1   =SUM(A1*B1)

Example: The amount you are thinking about borrowing for a short-term working capital loan is $5,000. The rate you were quoted was 8.0%. Simple interest is calculated as:

$$\$5,000 \times 8.0\% = \$400.00$$

---

3 Philip, M. (1945). A Note on Simple Interest. *National Mathematics Magazine*, 19(8), 414–417.

The spreadsheet values are:

A1 5,000

B1 .08

Basic math review: When multiplying by a percentage, convert the stated rate to decimal form. Shift the decimal two places to the left or divide by 100; either method produces the same result.

Formula: Percentage Conversion to Decimal

a) Decimal shift:

$$r.0\% = 0.0r \; decimal$$

b) Divide by 100:

$$r \div 100 = D$$

where: r = percentage rate

D = decimal equivalent

For example, for 8.0%:

$$8.0 \div 100 = 0.08$$

The recalculated decimal equivalent is used as the multiplier in the simple interest calculation. To make this calculation on a spreadsheet program, enter the following values:

A1 R

B1 =SUM(A1/100)

In the preceding example, A1 will contain the value 8.00 and this results in a result of 0.08.

Simple interest may be used for calculations in some loans, especially those due in one year or less. However, it is rare for most business loans. This calculation works as a sensible starting point for more complex interest calculations and for making comparisons between the stated, or nominal rate and the annual compound rate.

Daily Compound Interest

Most interest is compounded more than once per year. The most common rates are monthly and quarterly. Monthly compounding is equal to one-twelfth of the stated annual rate, used as each month's "periodic" rate. There are 12 periods for monthly compounding (and four for quarterly compounding). The formula for periodic rate is calculated by dividing the stated rate by the number of periods in the year.

Formula: Periodic Rate

$$R \div p = i$$

where: R = nominal interest rate
p = number of periods
i = periodic interest rate

On a spreadsheet program, enter the following values:
A1   R
B1   p
C1   =SUM(A1/B1)

Example: Your stated interest rate is 7.5%. Compounding takes place monthly, meaning there are 12 periods in the year. The periodic rate in this case is:

$$7.5\% \div 12 = .625\%$$

The spreadsheet values are:
A1   7.5
B1   12

Recalling the conversion formula, converting 7.5% to decimal form requires shifting the decimal two places to the left, or dividing by 100:

$$7.5 \div 100 = .075 \ decimal$$

Next, the decimal equivalent is divided by the number of periods. For monthly compounding, divide by 12:

$$0.075 \div 12 = .00625 \ decimal$$

You need to know the periodic rate to calculate interest for each applicable period, and also to figure out the compound annual rate. Starting with the method requiring the greatest amount of calculation—daily compounding, the number of periods per year is either 360 or 365. There are two methods used for daily compounding; Using 365 is the full year method and using 360 is known as the "banker's year" method.

To calculate daily compounding (using the 365-day method), first divide the full year's interest rate by 365. This produces the daily periodic rate.

Formula: Daily Periodic Rate (365 days)

$$R \div 365 = i$$

where: R = stated annual interest rate
$\quad i$ = periodic interest rate (365 days)

On a spreadsheet program, enter:
A1   stated annual interest rate (in decimal form)
B1   365
C1   =SUM(A1/B1)

Example: Your stated interest rate is 7.5% (or a decimal equivalent of 0.075). The method used for calculating interest is daily, based on the 365-days per year rate. The daily period rate is:

$$0.075 \div 365 = .0002055$$

Once you compute the daily rate, each day's interest is computed with a series of steps:
1.   Add '1' to the daily rate. This is the first day's multiplier for a debt:

$$.0002055 + 1 = 1.0002055$$

2.   Multiply the sum in the previous step by the amount of the debt. For example, if the amount borrowed is $8,000, the first day's debt (principal plus interest) interest is:

$$1.0002055 \times \$8,000.00 = \$8,001.64$$

3.   To calculate the subsequent day of the accumulated debt, multiply the answer above by the initial daily rate in step 1:

$$1.000205^2 \times \$8,001.64 = \$8,003.28$$

To calculate the effective interest for several days, you can use a shortcut method. Multiply the daily rate times the number of additional days, and then by the initial sum. For example, if you want to calculate the interest as of the fifth day, multiply the daily rate by itself four times (for days two through five, and then by the principal amount:

1.0002055 × 1.0002055 ×1.0002055 ×1.0002055 × 1.0002055 × $8,000.00 = $8,008.22

A shorthand version of this formula is:

$$1.0002055^5 \times \$8,000.00 = \$8,008.22$$

This can be verified by checking the steps for each of the five days:

Day	Rate	Total
		$8,000.00
1.	1.0002055	8,001.64
2.	1.0002055	8,003.29
3.	1.0002055	8,004.93
4.	1.0002055	8,006.58
5.	1.0002055	8,008.22

The formula for calculating daily compounding is:

Formula: Daily Compounding

$$((1 + (R \div i))^n) \times P = C$$

where: R = stated annual interest rate
i = periodic interest rate (365 days)
n = number of periods to be compounded
P = principal
C = compounded value

This series of calculations can also be placed on a worksheet and calculated using the formula feature. For spreadsheet programs, the following formulas are needed based on placement of information in named cells:

Daily Compounding

A1	annual interest rate divided by 365 = daily rate, plus 1 =SUM(i/365) + 1	
B1	principal amount	
C1	accumulated amount	=SUM (A1*B1)
A2		=A1
B2		=C1

To perform this Excel calculation, do the following:
1. Copy C1
2. Paste it to C2, so that C1 and C2 are the same
3. Copy the row A2, B2, and C2
4. Paste to row 3, columns A, B, and C
5. Repeat paste for each row

This process is carried forward to as many days as you need. You can use the exponent function on a calculator to make the calculation the fastest. If you do not have that feature, a fast shortcut for finding the effective daily rate for a large number of days is to multiply the daily rate (A3) by itself for as many days as needed (remembering that the initial sum is the first day). Thus, for the rate applicable on the 20th day, multiply the rate 19 more times by itself. You can do this on any calculator by entering the amount, then the 'x' button, and then the '=' button 19 times. In the case of the 7.5% annual (compounded daily) the 20th day's rate is:

$$1.0002055^{20} = 1.0041180$$

Next, multiple this by $8,000.00

$$1.0041180 \times \$8,000.00 = \$8,032.94$$

The outcome for 20 days based on the spreadsheet formula is summarized in Table 1.1.

Table 1.1: Daily compounding (365-day rate)
Deposit amount = $8,000.00
Interest rate = 7.5%
Term: 20 days

Day	A	B	C (value)
1	1.000205479	8,000.000	8001.64
2	1.000205479	8001.644	8003.29
3	1.000205479	8003.288	8004.93
4	1.000205479	8004.933	8006.58
5	1.000205479	8006.577	8008.22
6	1.000205479	8008.223	8009.87
7	1.000205479	8009.868	8011.51
8	1.000205479	8011.514	8013.16
9	1.000205479	8013.160	8014.81
10	1.000205479	8014.807	8016.45
11	1.000205479	8016.454	8018.10
12	1.000205479	8018.101	8019.75
13	1.000205479	8019.748	8021.40
14	1.000205479	8021.396	8023.04
15	1.000205479	8023.044	8024.69
16	1.000205479	8024.693	8026.34
17	1.000205479	8026.342	8027.99
18	1.000205479	8027.991	8029.64
19	1.000205479	8029.641	8031.29
20	1.000205479	8031.291	8032.94

The formula for calculating the daily debt (principal plus interest) is also called the accumulated value of '1.' A visual representation of the concept of compounding interest on a single deposit is shown in Figure 1.1.

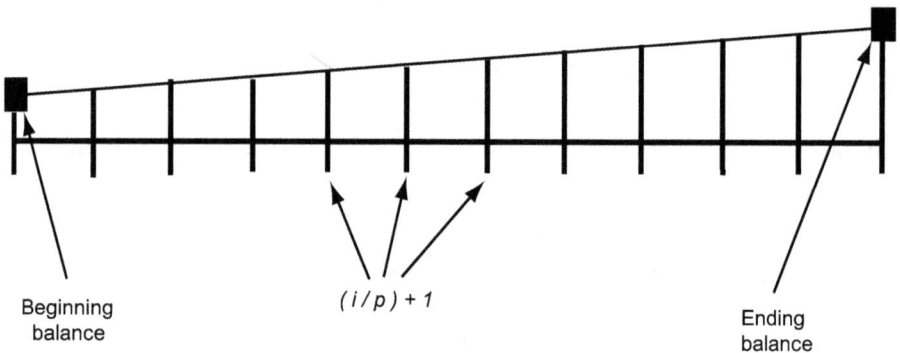

Beginning balance

$(i/p) + 1$

Ending balance

Figure 1.1: Accumulated value of a single deposit

Annual, Semi-annual and Quarterly Interest

The calculations for daily interest are far more complex than for annual (once per year), semi-annual (twice per year) and quarterly (four times per year).

Annual interest is not the same as simple interest, which involves no compounding. In the case of annual interest, compounding takes place only once per year. This is an unusual calculation and is likely to be found in private loans only, such as loans between a business and a family member. To calculate, the stated annual interest is compounded every year. For example, an $8,000 loan at 7.5% with annual compounding would involve a three-year calculation of a straight 8% at the end of each year:

Year	Interest	Total
		$8,000.00
1	$600.00	8,600.00
2	645.00	9,245.00
3	693.38	9,938.38

The calculation for annual compounding to arrive at the total debt is:

Formula: Annual Compounding

$$(i + 1)^x \times P = D$$

where:
i = annual interest rate
x = number of years
P = principal deposited
D = total debt as of the number of years

Example: Applying this formula to the example above, the formula is:

$$(.075 + 1)^3 \times \$8,000.00 = \$9,938.38$$

This can be summarized on a spreadsheet, with the following fields and values (note, the first calculation is the combined decimal equivalent of the interest rate plus 1; so, 7.5% would be equal to 0.075+ 1, or 1.075):

Annual compounding:

A1   annual interest rate plus 1 =SUM(i+1)
B1   principal amount
C1   accumulated amount   = SUM(A1*B1)
A2   =A1
B2   =C1

To perform this Excel calculation, do the following:
1. copy C1
2. paste to C2
3. copy A2, B2, and C2
4. paste to rows 3, columns 1, 2, and 3
5. repeat paste for each row

The outcome for three years using the spreadsheet program is shown in Table 1.2.

Table 1.2: Annual compounding
Deposit amount = $8,000.00
Interest rate = 7.5%
Term: 3 years

Year	A	B	C (Value)
1	1.075	8,000	8,600.00
2	1.075	8,600	9,245.00
3	1.075	9,245	9,938.38

Semi-annual interest is calculated twice per year. Using the same example as above, a three-year term requires six calculations rather than three, based on a periodic rate one-half of the annual rate. Since the annual rate is 7.5% (decimal equivalent .075), the semi-annual periodic rate is:

$$.075 \div 2 = .0375$$

The three-year accumulation of interest and principal in the case of semi-annual compounding is:

Period	Interest	Total
		$8,000.00
1	$300.00	8,300.00
2	311.25	8,611.25
3	322.92	8,934.17
4	335.03	9,269.20
5	347.60	9,616.80
6	360.63	9,977.43

This method produces an additional $39.05 above annual compounding. The more often the compounding is performed, the higher the compounding effect. The formula for semi-annual compounding is:

Formula: Semi-annual Compounding

$$((i \div 2) + 1)^x \times P = D$$

where: i = annual interest rate
x = number of semi-annual periods
P = principal deposited
D = total debt as of the number of periods

Example: Applying this formula to the above example, an $8,000 deposit with 7.5% semi-annual interest, for six periods (three years) results in a sum of:

$$((.075 \div 2) + 1)^6 \times \$8,000.00 = \$9,977.43$$

This can be summarized on a spreadsheet, with the following fields and values (note, the first calculation is the combined decimal equivalent of the interest rate plus 1; so, the half-year rate of 3.75% would be equal to 0.0375+ 1, or 1.0375):

Semi-annual compounding:

A1	annual interest rate plus 1	=SUM(i/2)+1
B1	principal amount	
C1	accumulated amount	=SUM(A1*B1)
A2		=A1
B2		=C1

To perform this Excel calculation, do the following:
1. Copy C1
2. Paste to C2
3. Copy A2, B2, and C2
4. Paste to rows 3, columns 1, 2, and 3
5. Repeat paste for each row

The calculations on the spreadsheet program are summarized for six periods in Table 1.3.

Table 1.3: Semi-annual compounding
Deposit amount = $8,000.00
Interest rate = 7.5%
Term: 3 years (6 semi-annual periods)

Period	A	B	C (Value)
1	1.0375	8,000.00	8,300.00
2	1.0375	8,300.00	8,611.25
3	1.0375	8,611.25	8,934.17
4	1.0375	8,934.17	9,269.20
5	1.0375	9,269.20	9,616.80
6	1.0375	9,616.80	9,977.43

A more common calculation of interest is based on quarterly compounding, or a calculation of interest four times per year. For example, over a three-year period, you would have 12 periods to calculate. Based on $8,000 at 7.5% compounded quarterly (1.875% per quarter):

Period	Interest	Total
		$8,000.00
1	$150.00	8,150.00
2	152.81	8,302.81
3	155.68	8,458.49
4	158.60	8,617.09
5	161.57	8,778.66
6	164.60	8,943.26
7	167.68	9,110.94
8	170.83	9,281.77
9	174.03	9,455.80
10	177.30	9,633.10
11	180.62	9,813.72
12	184.01	9,997.73

The formula for quarterly compounding is:

Formula: Quarterly Compounding

$$((i \div 4) + 1)^X \times P = D$$

where: i = annual interest rate
x = number of quarterly periods
P = principal deposited
D = total debt as of the number of periods

Example: Based on the same $8,000.00 deposit but using 7.5% and quarterly compounding, the formula for a three-year accumulation is:

$$((0.075 \div 4) + 1)^{12} \times \$8,000.00 = \$9,997.86$$

Quarterly compounding can be summarized on a spreadsheet, with the following fields and values (note, the first calculation is the combined decimal equivalent of the interest rate plus 1; so, the quarterly rate of 1.875% would be equal to 0.01875 + 1, or 1.01875):

Quarterly compounding:

A1	annual interest rate plus 1	=SUM(i/4) + 1
B1	principal amount	
C1	accumulated amount	= SUM(A1*B1)
A2		=A1
B2		=C1

To perform this Excel calculation, do the following:
1. Copy C1
2. Paste to C2
3. Copy A2, B2, and C2
4. Paste to rows 3, columns 1, 2, and 3

The formula based on this example as it appears on a spreadsheet is summarized in Table 1.4.

Table 1.4: Quarterly compounding
Deposit amount = $8,000.00
Interest rate = 7.5%
Term: 3 years (12 quarterly periods)

Period	A	B	C (Value)
1	1.01875	8,000.00	8,150.00
2	1.01875	8,150.00	8,302.81
3	1.01875	8,302.81	8,458.49
4	1.01875	8,458.49	8,617.09
5	1.01875	8,617.09	8,778.66
6	1.01875	8,778.66	8,943.26
7	1.01875	8,943.26	9,110.94
8	1.01875	9,110.94	9,281.77
9	1.01875	9,281.77	9,455.81
10	1.01875	9,455.81	9,633.10
11	1.01875	9,633.10	9,813.72
12	1.01875	9,813.72	9,997.73

Monthly Compounding

The best-known and most widely used calculation is monthly compounding. This method is used on virtually all mortgages and commercial loans. It involves dividing the nominal rate by 12 (months) and then applying the periodic rate to each month's outstanding balance.

For example, an $8,000 deposit at 7.5% compounded monthly involves 12 calculations per year, each at 1/12th of the nominal rate, or 0.00625%:

$$0.075 \div 12 = 0.00625$$

The formula for monthly compounding is:

Formula: Monthly Compounding

$$((i \div 12) + 1)^x \times P = D$$

where: i = annual interest rate
x = number of months
P = principal deposited
D = total debt as of the number of months

Over a one-year period, this involves 12 calculations, with each month's interest 0.00625. The first year's calculations are:

Month	Interest	Total
		$8,000.00
1	$50.00	8,050.00
2	50.31	8,100.31
3	50.63	8,150.94
4	50.94	8,201.88
5	51.26	8,253.14
6	51.58	8,304.73
7	51.90	8,356.63
8	52.23	8,408.86
9	52.56	8,461.42
10	52.88	8,514.30
11	53.21	8,567.51
12	53.55	8,621.06

Example: To apply the formula summarizing this calculation:

$$((0.075 \div 12) + 1)^{12} \times \$8,000.00 = \$8,6201.06$$

This can also be summarized on a spreadsheet with the same design as that shown for previous compounding methods. The annual rate, expressed in decimal form, is 0.075, and when divided by 12, produces the monthly rate of .00625. When 1 is added, the monthly multiplier is the result:

$$(0.075 \div 12) + 1 = 1.00625\%$$

The following is summary of the spreadsheet programming for monthly compounding:

Monthly compounding:

A1	annual interest rate plus 1	=SUM(i/12) + 1
B1	principal amount	
C1	accumulated amount	=SUM(A1*B1)
A2		=A1
B2		=C1

To perform this Excel calculation, do the following:
1. Copy C1
2. Paste to C2
3. Copy A2, B2, and C2
4. Paste to rows 3, columns 1, 2, and 3

A summary of the spreadsheet rows and columns is provided in Table 1.5.

Table 1.5: Monthly compounding
Deposit amount = $8,000.00
Interest rate = 7.5%
Term: 1 year (12 months)

Period	A	B	C (Value)
1	1.00625	8,000.00	8,050.00
2	1.00625	8,050.00	8,100.31
3	1.00625	8,100.31	8,150.94
4	1.00625	8,150.94	8,201.88
5	1.00625	8,201.88	8,253.14
6	1.00625	8,253.14	8,304.73
7	1.00625	8,304.73	8,356.63
8	1.00625	8,356.63	8,408.86
9	1.00625	8,408.86	8,461.42
10	1.00625	8,461.42	8,514.30
11	1.00625	8,514.30	8,567.51
12	1.00625	8,567.51	8,621.06

Although many free calculators are easily found online for calculations of interest, it is useful to be aware of how the calculations are performed. This improves your basic comprehension of the functions involved, and provides you with the ability to make calculations even when online calculators are not available at the moment.

Accumulated Value of a Series of Deposits

Calculating the accumulated value (also called 'future value') of a single deposit over various compounding methods is far easier than calculating how a fund grows with a *series* of deposits over time. This has widespread business applications. For example, if you are setting up a reserve with monthly deposits, or putting funds aside for expansion, you need to calculate how monthly deposits will grow over a period of months or years.

This calculation is going to vary based on the compounding method used. The more frequently the compounding occurs, the more interest will accumulate. However, the "periodic" interest rate used in the following calculations is going to vary, and this must be kept in mind. For example, a 7.5% annual nominal rate over three years contains different period rates and a different number of periods in the calculation:

Periodic interest rate (7.5% annual rate, three years):

Method	Periods	Periodic Rate
monthly	36	.00625
quarterly	12	.01875
semi-annually	6	.0375
annually	3	.075

In this formula, you need to calculate the growing effect of a series of deposits over time, which varies by the amount deposited, interest rate, overall time involved and compounding method. The formula is:

Formula: Accumulated Value of a Series of Deposits

$$D\,(((1 + R)^n - 1) \div R) = A$$

where: D = periodic deposit amount
R = periodic interest rate
n = number of periods
A= accumulated value

The accumulated value of a series of deposits is further illustrated in Figure 1.2.

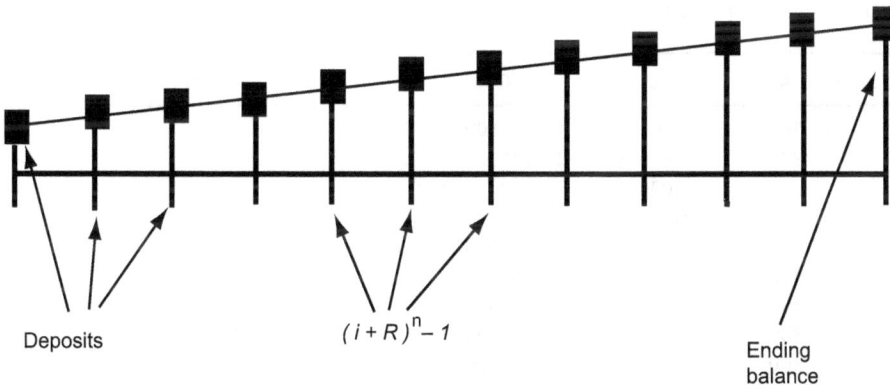

Deposits $(i + R)^n - 1$ Ending balance

Figure 1.2: Accumulated value of a series of deposits

Example: You are putting aside $300 per month for three years. Your fund is estimated to yield 7.5% compounded monthly. The formula is:

$$\$300.00\ (((1 + .00625)^{36} - 1) \div .00625) = A$$
$$\$300.00\ ((1.251446 - 1) \div .00625) = A$$
$$\$300.00\ (40.23136) = \$12,069.41$$

This formula can also be calculated using a spreadsheet formula program:

Accumulated value of a series of deposits:

A1	Periodic deposit amount	
B1	Periodic interest rate + 1	=SUM(1+R)
C1		=B1
B2		=B1
C2		=SUM(C1*B2)
	copy B and C, row 3	
	paste to subsequent B and C rows	
D36	(assuming 36 periods total)	=SUM(C36-1)/(C1-1)
E36		=SUM(D36*A1)

The outcome of the spreadsheet columns and rows is summarized in Table 1.6.

Table 1.6: Accumulated value of a series of deposits
Deposit amount = $300 per month
Interest rate = 7.5% compounded monthly
Term: 3 years (36 months)

Period	A	B	C	D	E (final value)
1	300	1.00625	1.006250000		
2	300	1.00625	1.012539063		
3	300	1.00625	1.018867432		
4	300	1.00625	1.025235353		
5	300	1.00625	1.031643074		
6	300	1.00625	1.038090843		
12	300	1.00625	1.077632599		
18	300	1.00625	1.118680533		
24	300	1.00625	1.161292018		
30	300	1.00625	1.205526610		
36	300	1.00625	1.251446136	40.2313817	12,069.41

The various methods of calculating return can be greatly simplified by using a few formulas, especially when also programmed onto a spreadsheet. In each situation, the variables provide great flexibility. So, the annual interest rate and compounding method are entered into the formula to replace previous results, ensuring ease of calculations even for extended time periods. In the spreadsheet program, the fields you copy can be pasted in multiple subsequent fields rather than one at a time. For example, in calculating daily compounding, once your formula has been entered, the daily outcome can be pasted in a single step to all 365 fields. This is what makes spreadsheet formulas effective and easy to use.

The next chapter introduces more advanced calculations, including present and future value. Although these formulas are more complex than calculating interest on single or periodic deposits, they can also be simplified and managed by programming formulas onto a spreadsheet.

Chapter 2
Present Value and Sinking Funds

The last chapter presented a few basic formulas. The accumulation of interest for a single deposit answers the question, "How much will a single sum grow over a period of time, given a known rate of interest and compounding method?" The accumulation of a series of deposits is a variation.

This chapter tackles more complex formulas and answers a different but equally important set of questions. The concepts of present value and sinking fund payments are based on knowing the desired result and determining how much you need to deposit today to achieve that balance. For managers, present value calculations are likely to be more pressing than tracking savings accounts. For example, if you need to set aside a reserve for future losses, save money to purchase equipment, or better manage cash flow by setting up a series of payments, these calculations are going to be a part of the process.

Present Value of a Single Deposit

The accumulation of money based on combinations of principal and interest are well understood, even if the calculations are not. On the other side of the interest calculation exercise is present value. This is needed to identify how to save to a target number in the future with a single deposit or a series of deposits, or how to pay off a loan within several months or years.

The first of these calculations is called the *present value of a single deposit*. It answers the question: "How much money do I need to deposit today, given a known rate of interest, compounding method, and time period, to accumulate a target amount in the future?"

Example: You plan to purchase a new piece of machinery in five years at an estimated cost of $8,000. You do not want to finance this purchase but prefer to save up for it. You need to know how much money you need to put away now to achieve that goal. You have identified a very successful mutual fund that has averaged a 7.5% return for the past three years. Based on the assumption that this record will be duplicated for the next five years, you need to know how much to put aside right now.

A calculation like this is always based on a series of assumptions. Among these are the most obvious one, the assumption that the mutual fund will achieve the same average rate of return in the future as it has in the past. Second is the assumption that you will reinvest all earnings (dividends, interest, and capital gains), because to accomplish your goal you will need the compounded rate of return. You also need to determine which compounding method is best to use. The advantage of formatting a

DOI 10.1515/9781547400638-002

calculation in a spreadsheet is that all the variables can be changed, and all outcomes will change accordingly. So even if you have entered monthly outcomes for five years, or 60 months, as soon as the variables change, you can also change the variables to keep the exercise realistic. If your mutual fund loses money in the first three months, your net fund will be worth less. By depositing additional funds (changing the variable) you can create a revised worksheet. If the fund's performance is unexpectedly positive, you can instantly see how much excess your five-year plan with create; this can be treated as a type of reserve against future cyclical downturns.

To calculate the present value of a single deposit, you need only to know what assumptions you will use: The annual average earnings rate, compounding method, and desired end-result fund. The formula:

Formula: Present Value of a Single Deposit

$$(1 \div (1 + (i \div p))^n) \times D = V$$

where: i = annual interest rate
p = number of periods in the compounding method
n = periods until the deposit amount is needed
D = end-result deposit
V = amount needed to be deposited today

This formula is not as complex as it seems at first glance. The value '1' is divided by a series of familiar calculations. The annual interest rate must be reduced to the periodic rate, and then the number of periods (normally months) calculated. A visual look at this is shown in Figure 2.1.

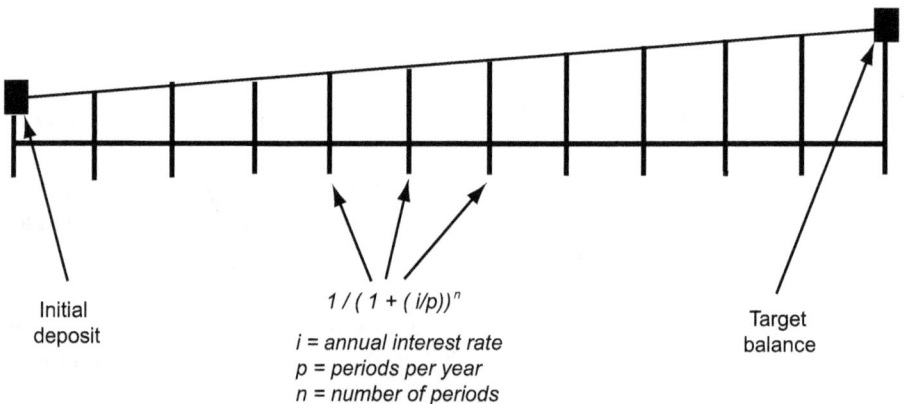

Initial deposit

$1 / (1 + (i/p))^n$
i = annual interest rate
p = periods per year
n = number of periods

Target balance

Figure 2.1: Present value of a single deposit

For monthly compounding, the annual rate must be divided by a 'p' of 12; for quarterly, 'p' is 4. The resulting factor (which will be a decimal value less than '1') is then multiplied by the amount of deposit needed at the end of the period to find the value of the single deposit you need today.

Example: You want to accumulate a fund of $8,000 in five years. Your assumed rate on an investment in a mutual fund is 7.5%, and you will calculate compounding monthly. How much do you need to place on deposit today? Apply the formula above to calculate. This can be reflected in a spreadsheet and then manipulated to check outcomes with varying assumptions. The spreadsheet formula:

Present value of a single deposit:

A1	periodic interest rate plus 1	=SUM(1+(i/p))
B1		=A1
A2		=A1
B2		=SUM(A2*B1)
	copy row 2, columns A and B	
	paste to subsequent rows	
C60	(assuming 60 months total)	=SUM(1/B60)
D60	deposit needed today	=SUM(C60*D)

This formula presents a scenario based on well understood assumptions. However, if you need to alter your assumptions, a copy-and-paste operation is easy to perform. For example, if you decide you need to meet your $8,000 target in 52 months instead of 60 months, you can copy cell C60 and D60 and paste into cells C52 and D52. The result is $5,786.07, the amount you need to deposit today rather than the 60-month outcome of $5,504.73.

The outcome of this calculation is summarized in Table 2.1.

Table 2.1: Present value of a single deposit
Final deposit amount = $8,000.00
Interest rate = 7.5%
Term: 60 months

Month	A	B	C	D
1	1.00625	1.00625		
2	1.00625	1.01254		
3	1.00625	1.01887		
4	1.00625	1.02524		
5	1.00625	1.03164		
6	1.00625	1.03809		
7	1.00625	1.04458		
8	1.00625	1.05111		

Table 2.1: (continued)

Month	A	B	C	D
9	1.00625	1.05768		
10	1.00625	1.06429		
15	1.00625	1.09796		
20	1.00625	1.13271		
25	1.00625	1.16855		
30	1.00625	1.20553		
35	1.00625	1.24367		
40	1.00625	1.28303		
45	1.00625	1.32363		
50	1.00625	1.36551		
55	1.00625	1.40872		
60	1.00625	1.45329	0.688091824	5,504.73

Sinking Fund Payments and Present Value Per Period

Making a single deposit at the beginning of a period is rarely a practical idea. Managers are likely to make a series of monthly deposits over a period of months to accumulate a known future value. The "sinking fund payment" is a calculation of a series based on known assumptions. It answers the question, "How much money do I need to deposit per month, given a known rate of interest, compounding method, and time period, to accumulate a target amount in the future?"

As you would expect the formula for this calculation is more complex than for a single deposit. It involves figuring out compound interest over many months and based on a series of recurring payments into a fund. The formula for sinking fund payments:

Formula: Sinking Fund Payments

$$D \times ((i \div p) \div ((1 + (i \div p))^n - 1)) = V$$

where: D = target deposit
i = annual interest rate
p = number of periods in the compounding method
n = periods until the deposit amount is needed
V = amount of periodic deposits required

This formula is complex because it requires a double level of division functions. However, it is simplified by placement on a spreadsheet program. In addition, it is better comprehended when viewed; refer to Figure 2.2 for a visual summary of this formula.

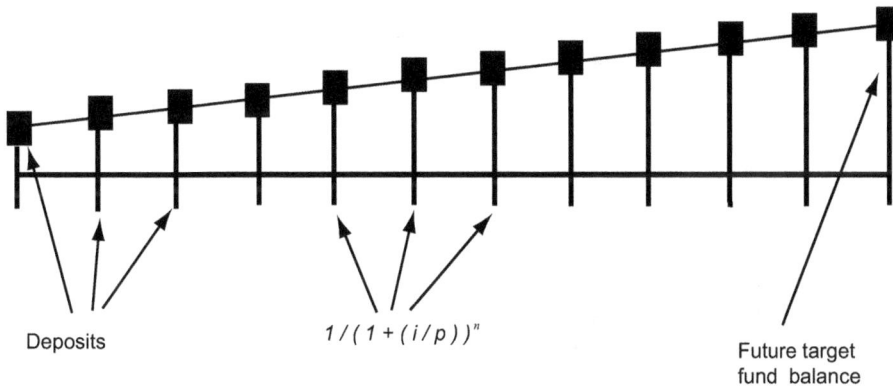

Figure 2.2: Sinking fund payments

Example: You want to accumulate $8,000 in 60 months. Your assumptions are that you will earn 7.5% per month, compounded monthly (12 times per year). You need to know the value of a series of deposits (sinking fund payments) required for the next 60 months based on these assumptions. The spreadsheet formula:

Sinking fund payments:

A1	periodic interest rate plus 1	=SUM(1+(i/p))
B1		=A1
A2		=A1
B2		=SUM(A2*B1)
	1. copy row 2, columns A and B	
	2. paste to subsequent rows	
C60		=SUM(A1–)/(B60–1)
D60	deposits needed	=SUM(C60*D)

In the example, the values of these abbreviations are:

$i = 0.075$ (7.5% in decimal form)

$p = 12$ (periods required for monthly compounding)

$D = \$8,000$ (the fund needed at the end of 60 months)

The required monthly deposit is $110.30 for 60 months to create a fund worth $8,000.00. The calculated spreadsheet results are summarized in Table 2.2.

Table 2.2: Sinking fund payments
Final deposit amount = $8,000.00
Interest rate = 7.5%
Term: 60 months

Month	A	B	C	D
1	1.00625	1.00625		
2	1.00625	1.01254		
3	1.00625	1.01887		
4	1.00625	1.02524		
5	1.00625	1.03164		
6	1.00625	1.03809		
7	1.00625	1.04458		
8	1.00625	1.05111		
9	1.00625	1.05768		
10	1.00625	1.06429		
15	1.00625	1.09796		
20	1.00625	1.13271		
25	1.00625	1.16855		
30	1.00625	1.20553		
35	1.00625	1.24367		
40	1.00625	1.28303		
45	1.00625	1.32363		
50	1.00625	1.36551		
55	1.00625	1.40872		
60	1.00625	1.45329	0.013787949	110.30

In calculating the present value per period, the question being answered is: "How much do I need to deposit today to make a series of withdrawals in a specific time, assuming a rate of interest, compounding method, and number of months?"

The amount needed today to fall to zero at the end of the period, based on the amount of each withdrawal, the number of months, interest rate, and compounding method. For example, assume that you want to set up a reserve and draw $300.00 against it each month for five years. This kind of calculation occurs when you are going through business expansion or retiring obligations based on fixed monthly payments. The formula for the present value per period is:

Formula: Present Value Per Period

$$W \times (1 \div (1 + (i \div p)^n)) \div (i \div p) = D$$

where: W = periodic withdrawal amounts
i = annual interest rate
p = number of periods in the compounding method
n = periods until the deposit amount is depleted
D = initial deposit required

This formula produces the dollar amount needed to be deposited at the beginning of the term, based on monthly withdrawals, interest rate, compounding method, and number of months. In the example, the desired withdrawal amount was $300 per month for five years (60 months), based on monthly compounding of a 7.5% annual rate. The outcome can be placed on a spreadsheet as shown below.

Present Value per period:

A1	periodic interest rate plus 1	=SUM(1+(i/p))
B1		=A1
A2		=A1
B2		=SUM(A2*B1)

To perform this Excel calculation, do the following:
1. copy row 2, columns A and B
2. paste to subsequent rows

C60	=SUM(1-(1/B60))	
D60	dollar amount to be deposited	=SUM(C60/(A1–1))*W

The variables in the preceding example are:
i = .075 (decimal equivalent of 7.5%
p = 12 (periods in monthly compounding)
W = $300.00 monthly withdrawal amount.

This is also summarized in Figure 2.3.

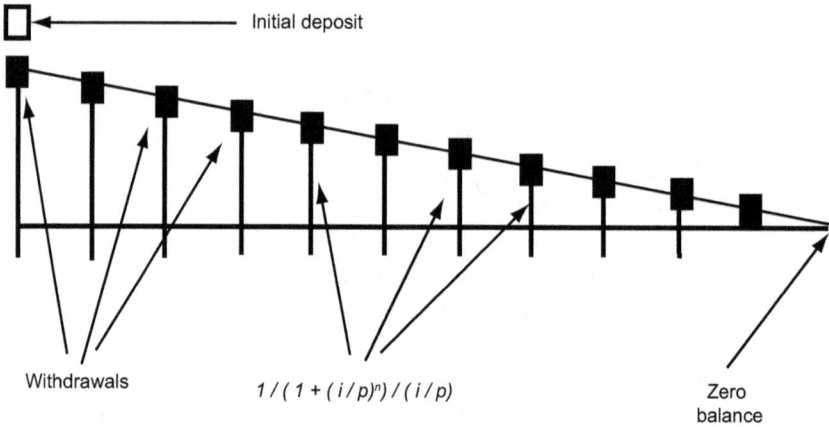

Figure 2.3: Present value per period

Based on this calculation, you need to deposit $14,971.59 to withdraw $300 per month for 60 months and deplete the fund. The calculation is summarized based on the spreadsheet, in Table 2.3.

Table 2.3: Present value per period
Monthly withdrawal amount = $300.00
Interest rate = 7.5%
Term: 60 months

Month	A	B	C	D
1	1.00625	1.00625		
2	1.00625	1.01254		
3	1.00625	1.01887		
4	1.00625	1.02524		
5	1.00625	1.03164		
6	1.00625	1.03809		
7	1.00625	1.04458		
8	1.00625	1.05111		
9	1.00625	1.05768		
10	1.00625	1.06429		
15	1.00625	1.09796		
20	1.00625	1.13271		
25	1.00625	1.16855		
30	1.00625	1.20553		
35	1.00625	1.24367		
40	1.00625	1.28303		
45	1.00625	1.32363		
50	1.00625	1.36551		
55	1.00625	1.40872		
60	1.00625	1.45329	0.311908176	14,971.59

Loan Amortization

Yet another variety of the present value formula is loan amortization. This is the procedure used to retire any long-term debt and is best known for mortgages.

Example: You purchase the building where your company is based and make payments over a period of years, and the loan terms require you to amortize (pay down) the loan each month. Every payment consists of principal and interest; the interest is calculated based on the outstanding balance. For this reason, interest is high during the early years of a repayment schedule and declines over time. In a 30-year mortgage at 7.5%, for example, it takes 22 out of 30 years to pay off one-half of the debt, with the remaining 50% paid during the last eight years.

The reasons for the slow decline in mortgage principal all have to do with interest. To calculate loan amortization, the formula incorporates the formula for the present value per period in a more involved calculation. The present value per period introduced in the preceding section was:

$$W \times (1 \div (1 + (i \div p))^n) \div (i \div p) = D$$

where: W = periodic withdrawal amounts
i = annual interest rate
p = number of periods in the compounding method
n = periods until the deposit amount is depleted
D = initial deposit required

This is slightly modified as part of the loan amortization formula. The initial value of 'W' is going to be '1' so it is not needed as part of this formula (since multiplying an equation by '1' does not change its value). The loan amortization formula results in the required monthly payment to retire the full loan balance. The "present value of 1 per period" formula is:

$$(1 \div (1 + (i \div p))^n) \div (i \div p)$$

Formula: Loan Amortization (Summarized)

$$L \times (R \times P^n \div ((P^n) - 1) = A$$

where: L = original balance of the loan
R = periodic interest rate (annual rate divided by periods per year)
P = present value of 1
n = number of periods (usually months)
A = required payment per period

This formula and its outcome are summarized in Figure 2.4.

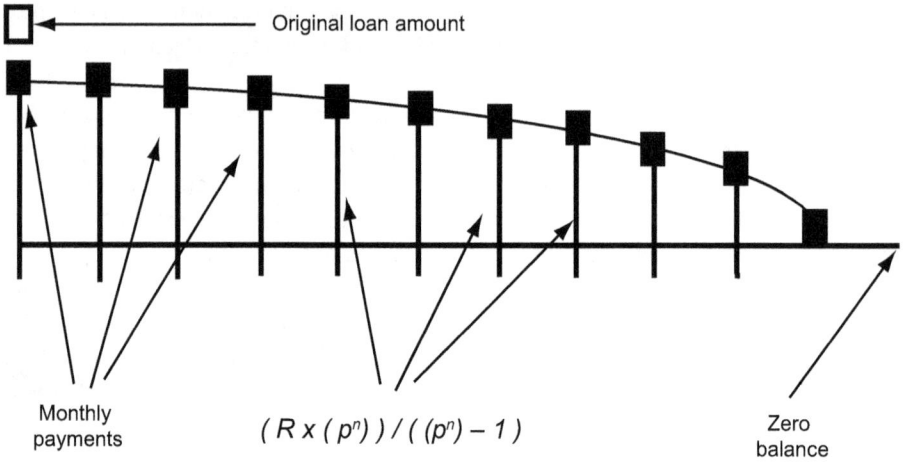

Figure 2.4: Loan amortization

This formula can also be summarized on a spreadsheet program using the following equations:

Loan amortization:

A1	periodic interest rate plus 1	=SUM(1+(i/p))
B1		=A1
A2		=A1
B2		=SUM(A2*B1)

To perform this Excel calculation, do the following:
1. copy row 2, columns A and B
2. paste to subsequent rows

C360	(assuming a 30-year mortgage term, or a total of 360 months)	=SUM(A360−1)*B360
D360	payment required	+SUM((C360/(B360−1))*L

In this example, the outcome for 360 months, based on 7.5% compounded monthly and for a $10,000 loan, is a required payment of $69.92 per month. For example, if you contracted a 30-year mortgage on your home to generate working capital, you could borrow the $10,000 on a 30-year contract with payments of $69.92 per month. Although this is a relatively small amount for a 30-year loan, a larger amount can be calculated easily by changing the spreadsheet variables. The calculations are summarized in Table 2.4.

Table 2.4: Loan amortization
Amount borrowed = $10,000
Interest rate = 7.5%
Term: 360 months

Month	A	B	C	D
1	1.00625	1.00625		
2	1.00625	1.01254		
3	1.00625	1.01887		
4	1.00625	1.02524		
5	1.00625	1.03164		
10	1.00625	1.06429		
20	1.00625	1.13271		
40	1.00625	1.28303		
60	1.00625	1.45329		
80	1.00625	1.64616		
100	1.00625	1.86462		
120	1.00625	2.11206		
140	1.00625	2.39235		
160	1.00625	2.70984		
180	1.00625	3.06945		
200	1.00625	3.47679		
220	1.00625	3.93819		
240	1.00625	4.46082		
260	1.00625	5.05280		
280	1.00625	5.72335		
300	1.00625	6.48288		
320	1.00625	7.34321		
340	1.00625	8.31771		
360	1.00625	9.42153	0.058884587	69.92

Loan amortization payments can also be found in interest tables and online, including free online calculators.

When you are making payments on a loan, you also need to track several features to determine the status of the loan (and to ensure that the lender is making calculations accurately). These include the balance forward, amount assigned to interest and principal, and the percentage remaining on the loan. Four separate calculations are involved, which can be reduced to a single spreadsheet program. The four calculations produce each month's interest, each month's principal, a new balance forward, and the percentage of remaining balance on the loan. These four formulas are:

Formula: Monthly Interest on a Loan

$$I = B \times (i \div p)$$

where: I = monthly interest
B = balance forward
i = annual interest rate
p = number of periods in the year (usually 12 months)

On a spreadsheet program:
A1   balance forward
B1   =SUM(interest rate/12)
C1   =SUM(A1*B1)

Formula: Monthly Principal on a Loan

$$P = M - I$$

where: P = monthly principal
M = monthly payment
I = monthly interest

On a spreadsheet program:
A1   M
B1   I
C1   =SUM(A1-B1)

Formula: New Balance Forward on a Loan

$$N = B - P$$

where: N = New monthly balance forward
B = balance forward
P = monthly principal

On a spreadsheet program:
A1   B
B1   P
C1   =SUM(A1-B1)

Formula: Remaining Balance Percentage on a Loan

$$R = N \div L$$

where: R = remaining balance percentage
    N = New monthly balance forward
    L = original loan amount

On a spreadsheet program:

A1	N
B1	L
C1	=SUM(A1/B1)

These four steps are easily understood when their purpose and outcome are summarized:

1. *Monthly interest on a loan:* This is a calculation in which the periodic rate (1/12th of the annual rate) is applied against the balance forward to calculate the current month's interest.
2. *Monthly principal on a loan:* The total monthly payment is reduced by the amount of interest, to isolate the amount going to pay down the loan.
3. *New balance forward on a loan:* The balance forward is reduced by the portion of the current month's payment going to principal.
4. *Remaining balance percentage on a loan:* This is the percentage of the original loan remaining due after applying the current month's principal.

All these steps can be placed on a spreadsheet formula to produce a worksheet for the entire loan period or for a portion of it. The formulas are:

Loan payments and balances:

		B
A1	loan amount ($10,000 in the example used)	
B1	monthly interest (7.5% per year, compounded monthly)	
		=SUM(A1*(i/p))
C1	monthly principal ($69.22 for 30 years in the example)	=SUM(M-B1)
D1	new balance forward (previous less principal)	=SUM(A1-C1)
E1	remaining balance percentage	=SUM(D1/A1)
A2		=D1

To perform this Excel calculation, do the following:
1. copy B1, C1, D1, E1
2. paste to B2, C2, D2, E2
3. copy columns A through E, row 2
4. paste to All subsequent cells (for as long as calculation is to be performed, such as 360 for a 30-year loan)

The formulas and outcome are summarized in Table 2.5.

Table 2.5: Loan payments and balances
Amount borrowed = $10,000
Interest rate = 7.5%
Term: 360 months

Month	Balance forward A	Interest B	Principal C	New balance forward D	Remaining balance E
1	10,000.00	62.50	7.43	9,992.57	99.93%
2	9,992.57	62.45	7.48	9,985.09	99.85
3	9,985.09	62.41	7.52	9,977.57	99.78
4	9,977.57	62.36	7.57	9,970.00	99.70
5	9,970.00	62.31	7.62	9,962.38	99.62
6	9,962.38	62.26	7.67	9,954.72	99.55
7	9,954.72	62.22	7.71	9,947.00	99.47
8	9,947.00	62.17	7.76	9,939.24	99.39
9	9,939.24	62.12	7.81	9,931.43	99.31
10	9,931.43	62.07	7.86	9,923.58	99.24
40	9,673.01	60.46	9.47	9,663.54	96.64
80	9,244.00	57.78	12.15	9,231.85	92.32
120	8,693.57	54.33	15.60	8,677.98	86.78
160	7,987.36	49.92	20.01	7,967.35	79.67
200	7,081.26	44.26	25.67	7,055.59	70.56
240	5,918.72	36.99	32.94	5,885.78	58.86
280	4,427.14	27.67	42.26	4,384.88	43.85
320	2,513.41	15.71	54.22	2,459.19	24.59
360	58.05	0.36	58.05	0.00	0.00

Reading Loan Amortization Tables

Rather than calculating the required monthly payment for every instance, you can use one of many free online calculators or refer to a look of loan amortization tables. These provide you with the interest rate, repayment term in years, the amount borrowed and the required monthly payment. A typical summarized loan amortization schedule is shown in Figure 2.5.

interest   years to
rate   repay

7.5%

TERM	5 YRS	10 YRS	15 YRS	20 YRS	25 YRS	30 YRS
AMOUNT						
500	10.02	5.94	4.64	4.03	3.70	3.50
1,000	20.04	11.88	9.28	8.06	7.39	7.00
2,000	40.08	23.75	18.55	16.12	14.78	13.99
3,000	60.12	35.62	27.82	24.17	22.17	20.98
4,000	96.72	47.49	37.09	32.23	29.56	27.97
5,000	100.19	59.36	46.36	40.28	36.95	34.97
10,000	200.38	118.71	92.71	80.56	73.90	69.93
15,000	300.57	178.06	139.06	120.84	110.85	104.89
20,000	400.76	237.41	185.41	161.12	147.80	139.85
25,000	500.45	296.76	231.76	201.40	184.75	174.81
30,000	601.14	356.11	278.11	241.68	221.70	209.77
35,000	701.33	415.46	324.46	281.96	258.65	244.73
40,000	801.52	476.81	370.81	322.24	295.60	279.69
45,000	901.71	534.16	417.16	362.52	322.55	314.65
50,000	1,001.90	593.51	463.51	402.80	369.50	349.61
55,000	1,102.09	652.86	509.86	443.08	406.45	384.57
60,000	1,202.28	712.22	556.21	483.36	443.40	419.53
65,000	1,302.47	771.57	602.56	523.64	480.35	454.49
70,000	1,402.66	830.92	648.91	563.92	517.30	489.46
75,000	1,502.85	890.27	695.26	604.20	554.25	524.42

amount borrowed

monthly payment

Figure 2.5: Loan amortization table

You can also estimate the amount of a required payment within pennies for loan amounts not shown on a table. For example, the monthly payment over 30 years for borrowing $500 is $3.50 per month at 7.5%, as the table reveals. For a loan of $1,000, the monthly payment is twice as much, or $7.00 per month. From this you can deduce that a loan halfway in between, for $750, will require a payment that is the average of these two:

$$(\$3.50 + \$7.00) \div 2 = \$5.25$$

This can be further tested by reducing the required payment for a $75,000 loan by 1/100th. The table shows that a $75,000 loan requires a monthly payment of 524.42. The calculation:

amount borrowed: $75,000 ÷ 100 = $750
monthly payment: $524.42 ÷ 100 = $5.24

Another way you can use tables for estimates is to calculate payments for interest rates not shown on the tables. For example, if tables are reported in half-point increments, you can calculate the approximate payment by averaging monthly payments for two known interest rates. For example, you know that at 7.5%, the monthly payment for a $10,000 loan over 30 years is $69.93 according to the table. (Earlier in the chapter, the calculated payment was $69.92. The difference comes from the specific method used in rounding.)

Referring to the 8.0% table, the monthly payment for borrowing $10,000 for 30 years is listed as $73.38. What is the monthly payment for 7.75%. To calculate, add the two known payments together and divide by two:

Interest rate: (7.50% + 8.00%) ÷ 2 = 7.75%
Monthly payment: ($ 69.93 + $73.38) ÷ 2 = $71.66

According to the actual 7.75% table, the actual payment required is $71.65.

These calculations make the point that payment requirements can be estimated based on variations of the rate, number of years and amount borrowed. In addition, by changing the variables in the spreadsheet program, you can determine the monthly payment for any interest rate, amount borrowed, or repayment term.

If you are working with two or more different loans and you want to find the average interest rate, it is not accurate to simply add them together and divide by the number of loans. The outcome is going to be distorted when the amount borrowed is different. To determine the average rate, you need to calculate the weighted average; this means that greater weight is given to larger amounts borrowed, and less to smaller amounts.

Example: You have two loans outstanding. The first was a $10,000 loan at 7.5%, and the second was an $18,000 loan at 12.0%. What is the weighted average of these loans? To calculate, you cannot add together the two rates and then divide by two (producing an average interest rate of 9.75%). Because the amount borrowed at the higher rate is more than the borrowed amount at the lower rate, you know that the weighted average must be greater. To weight the average, add the amounts borrowed and then use the total as the denominator in a fraction for the sum.

Basic math review: A fraction consists of the top number, or numerator; and the bottom number, or denominator. To multiply a whole number by a fraction, multiply the number by the numerator and then divide by the denominator. For example, if you want to find 3/8ths of 47:

$$(3 \times 47) \div 8 = 17.625$$

Proof: To prove this outcome, calculate in the reverse order: Divide 17.625 by 3, then multiply by 8:

$$(17.625 \div 3) \times 8 = 47$$

Applying the use of the fraction to the calculation for weighted average, first add together the two loan amounts:

$$\$10,000 + \$18,000 = \$28,000$$

Since fractions are always the same when reduced in size, these expressions can be shown as 10,000/28,000 and 18,000/28,000 or simply as 10/28 and 18/28. They can also be simplified down to their lowest form, or 5/14 and 9/14.

Basic math review: Fractions can be reduced to their lowest common denominator (LCD) to simplify functions using those fractions. The LCD is defined as the smallest equivalent fraction. For example, the fraction 6/10 is equivalent to the LCD of 3/5. To calculate, divide both sides of the fraction by a divisor that goes into both. In the case of 10,000/28,000, start by eliminating the zeros; next determine what divisor goes into both 10 and 18. These can both be divided by 2, creating the equivalent of 10/18 in the form of 5/9.

The fractions 5/14 and 9/14 are the same as 10,000/28000 and 18,000/28,000. Using the lowest form of the fraction, you can find the weighted average by multiplying the known interest rates by the reduced fractions for the loan amounts. You know that 5/14 of the loan ($10,000) is being paid at 7.5% and that 9/14 is being paid at 12.0%, so the weighted average is:

$$(7.5\% \times 5/14) + (12.0\% \times 9/14) =$$
$$(2.679) + (7.714) = 10.393\%$$

The weighted average of these two loans is 10.393%. The formula for weighted average is:

Formula: Weighted Average

$$(I_1 \times (L_1 \div L_t)) + (I_2 \times (L_2 \div L_t)) = W$$

where: I_1 = interest rate, loan '1'
L_1 = borrowed amount, loan '1'
L_t = total of amounts borrowed
I_2 = interest rate, loan '2'
L_2 = borrowed amount, loan '2'
W = weighted average

On a spreadsheet program, enter the following values:

A1	loan amount '1'	L_1
B1	loan amount '2'	L_2
C1		=SUM(A1+B1)
A2	interest rate, loan '1'	I_1
B2	interest rate, loan '2'	I_2
A3		=SUM(A2*(A1/C1))

B3	=SUM(B2*(B1/C1))
C3 weighted average	=SUM(A3+B3)

The previous example, based on a $10,000 loan at 7.5% and an $18,000 loan at 12.0%, loans out based on this formula to:

	A	B	C
1.	10,000	18,000	28,000
2.	.075	0.12	
3.	2.679	7.714	10.393

Calculating weighted average is not a complicated process, if the relative role of each loan is expressed in fractional form. The total amount borrowed is always the denominator, and each loan represents the numerator.

Calculating the interest cost of money is a well-known version of how interest works. The flip side of the same series of equations is the calculation of *return* in various forms. Return (also called yield) is the outcome calculated on revenues, cash, or invested capital. The next chapter provides calculations and examples for many versions of return.

Chapter 3
Rates of Return

Every manager faces the need to calculate returns on a variety of activities. The most pressing of these is the financial return, that all-important analysis of profitability. The profit, analyzed in comparison to revenues, is the universal test of how well you perform.

Besides the net return, there are other tests that are equally important as measurements of financial performance. These include return on cash or capital, dividend yield and comparisons between equity and debt. The return (or yield) is universally recognized for many financial and cash flow-related tests; the range of applications makes the analysis of "return" complex because it has so many definitions. This chapter analyzes and explains the various forms of return and yield in specific categories: revenue and equity, cash flow, capital, debt, and investment.

Return on Revenue and Equity

Within the range of return on revenue and equity, an array of financial calculations comes into play. Every manager, even those not involved directly in accounting and financial reporting, will end up being involved in the monitoring of business activity and its monitoring, which is invariably expressed in terms of return. Every manager will be judged based on how consistently a return is generated and by how much growth is generated. This is the primary method for making judgments about a manager's success or failure.

Net return (or, as more commonly expressed, *earnings per share*) is one of the most significant calculations you can perform. This is not restricted to the realm of the accounting department or treasurer's office; earnings levels and future earnings growth define the health of the organization in many ways:

> Growth of earnings per share determines future funds available for reinvestment in the corporation and for payment of dividends. It also influences future debt and equity financing, helps set cash income returns to shareholders and partly affects future changes of prices of equity securities. For these reasons accurate estimates of future growth rates of earnings per share are very important to the investor, the financial analyst, the corporate financial manager and the student of corporate finance.[1]

---

1 Murphy, J. (1967). Return on Equity Capital, Dividend Payout and Growth of Earnings per Share. *Financial Analysts Journal*, 23(3), 91–93.

DOI 10.1515/9781547400638-003

Even beyond the calculated earnings growth, other related calculations must be addressed. Because there are so many variations on *return*, it is crucial to make clear distinctions. The two most important distinctions are return on revenue and return on equity. Revenue, also called sales or gross receipts, is simply the percentage of net profit. This is one of the most widely used ratios. In comparison, return on equity is the percentage of net profit to net worth. The equity (net worth) is also called shareholders' equity or stockholders' equity in corporations.

Net return on revenue is easy to calculate; however, there are variations in which numbers are used in the calculation and this confuses comparisons between companies. Some "net return" calculations are based on net operating profit (profit before non-operating income or expenses such as interest, capital gains, foreign exchange profit or loss and tax liabilities). These non-operating values can be quite large, affecting the change between operating profit and overall net. Other calculations are based on pre-tax profit, even though a tax liability can be substantial; the after-tax net profit is the "net net" or the bottom line.

The selection of one number over another should be based on ensuring accurate year-to-year tracking. The non-operating values can distort the profit in years when they are quite high.

Example: A company's operating profit is $5,000 on total sales of $100,000 (5%). In the same year, a large capital gain of $15,000 was also booked when the company sold equipment at a profit. Total pre-tax profit was $20,000 (20%). The current year's tax liability was $6,000, bringing after-tax profits down to $14,000 (14%). The selection of one of these forms of profit drastically affects "net return" calculations:

net operating profit	5%
net pre-tax profit	20%
net after-tax profit	14%

Because "net profit" can mean different values based on how it is selected, you need to ensure that any comparisons between two different periods or two different companies are truly comparable. Otherwise, any analysis is going to be inaccurate.

Formula: Net Return

$$P \div R = N$$

where: P = net profit
R = revenue
N = net return

On a spreadsheet program, enter the following values:
A1 P
B1 R
C1 =SUM(A1/B1)

For example, if net profit is $427,600 and revenue is $5,342,400, the spreadsheet entries are:
A1 427,600
B1 5,342,400

Applying the formula, cell C1 is 0.0800 (8.0%).

Net return is important because it reveals how successfully the company managed costs and expenses during the year. You want to see profits rise with revenue, but just looking at the numbers can be deceptive, which is why the use of formulas is essential. One negative trend to look out for is a falling net return while revenues are rising. For example, look at a record of several years' revenues and net profits:

Year	Net profit	Revenues
1	$427,600	$ 5,342,400
2	461,000	7,616,300
3	494,500	9,827,900
4	507,800	12,625,000
5	527,200	17,570,400

At first glance this record looks impressive. Revenues have been rising substantially every year along with the dollar value of net profits:

Year	Net profit	Revenues	Net return
1	$427,600	$ 5,342,400	8.0%
2	461,000	7,616,300	6.1
3	494,500	9,827,900	5.0
4	07,800	12,625,000	4.0
5	527,200	14,310,400	3.7

Although the dollar value of both revenue and net profits rose over five years, the net return fell by more than 50%. This could be caused by a higher cost of doing business at a higher volume, or it could be a sign that costs and expenses are outpacing the growth in revenue. This is worth investigation to determine the answer. The dollar value analyzed without the net return does not reveal the trend, and the trend does not always fully explain the underlying cause for the trend. However, whatever further analysis reveals, it is a worthwhile pursuit.

Return on equity is a similar calculation. Net profit is divided by equity. In the case of net return, both net profit and revenue cover the same period (for example, one year's profits are divided by one year's total revenues). However, in the case of return on equity, the period's net profits are divided by the ending balance for the year (or quarter). In the first instance, both values come from the income statement; in the second, the income statement value (representing a period's activity) is divided by a balance sheet value (representing the net worth at the end of the period).

Formula: Return on Equity

$$P \div E = N$$

      where: P = net profit
              E = equity (net worth)
              N = return on equity

On a spreadsheet program, enter the following values:
    A1 P
    B1 E
    C1 =SUM(A1/B1)

Example: Your company reported net profit this year of $428,000. The equity (net worth) as of the end of the year was $5,226,000. Return on equity is:

$$\$428,000 \div \$5,226,000 = 8.2\%$$

The spreadsheet entries confirm this:
    A1 428,000
    B1 5,226,400

Applying the formula, cell C1 is 0.0819 (8.2%).

Return on equity shows how effectively your company has been able to put its resources to work in generating profits. This formula is not entirely accurate if the value of equity has changed during the year. For example, if additional stock has been created and sold, both beginning and ending balances of "net equity" are not representative of the entire year. The net income is income for the whole year, but if equity changes, then its value must be based on the average for the full year.

Example: Your company began the year with net worth of $4,000,000. On May 1, additional shares of common stock were sold and valued at $1,000,000. On September 1, an additional offering of $500,000 was made. On November 1, the company bought $274,000 worth of shares and retired them. The year-end balance of net worth was $5,226,000. The value for each month was:

January	$4,000,000
February	4,000,000
March	4,000,000
April	4,000,000
May	5,000,000
June	5,000,000
July	5,000,000
August	5,000,000
September	5,500,000
October	5,500,000
November	5,226,000
December	5,226,000

To find the average, add up the values and then divide by 12 (months).

Formula: Simple Average

$$(V_1 + V_2 + ... \; V_n) \div n = A$$

where: V = value

$_{1, 2}$ = field number

$_n$ = last number in the field

A = average

Expressing the same formula in a simplified form, add up the values for each month and then divide by 12:

$$57,452,000 \div 12 = 4,787,667$$

On a spreadsheet program, enter the following values (based on assumed 12 values:

A1	V_1
A2	V_2
Final cell (A12)	V_n
B12	=SUM(A1:A12)/12

The outcome using the 12 values in the example is shown in Table 3.1.

Table 3.1: Simple average

Month	A	B
January	4,000,000	
February	4,000,000	
March	4,000,000	
April	4,000,000	
May	5,000,000	
June	5,000,000	
July	5,000,000	
August	5,000,000	
September	5,500,000	
October	5,500,000	
November	5,226,000	
December	5,226,000	4,787.667

Note: Alternatively, you could use the Average function and this could be calculated in Excel as AVERAGE(A1:A12).

In the case of the changed value of equity throughout the year, this formula can be proven by multiplying the balance by a fraction for each portion of the year (for example, four months is equal to the fraction 4/12):

$$
\begin{array}{lll}
4/12 \times \$4,000,000 = & \$1,333,333 \\
4/12 \times 5,000,000 = & 1,666,667 \\
2/12 \times 5,500,000 = & 916,667 \\
2/12 \times 5,226,000 = & \underline{871,000} \\
\text{Total} & \$4,787,667
\end{array}
$$

This average is used in the calculation of return on equity to make it representative of the entire year. The average net worth must be used because both the beginning and ending balances for the year distort the true picture. The adjusted return on equity is:

$$\$428,000 \div \$4,787,667 = 8.9\%$$

The formula is adjusted in one specific circumstance. When a corporation has issued redeemable preferred stock (a special class of equity that acts more like a bond than like equity), the formula should exclude its value. This formula produces the *net* return on equity.

Formula: Net Return on Equity

$$P \div (E - S) = N$$

where: P = net profit
E = equity (net worth)
S = Redeemable preferred stock
N = net return on equity

On a spreadsheet program, enter the following values:
A1   P
B1   E
C1   S
D1   =SUM(A1/(B1-C1))

Example: Your company has reported net income of $428,000 and had an average equity value for the year of $4,787,667. Return on equity was calculated at 8.8%. However, equity included $300,000 in redeemable preferred stock. On the spreadsheet, enter the values:
A1   428,000
B1   4,787,667
C1   300,000
D1   =SUM(A1/(B1-C1))

The return on net equity is:

$$\$428,000 \div (\$4,787,667 - \$300,000) = 9.5\%$$

Although the formulas for return on revenue and return on equity are simple, finding the right values is not always as easy. Ensuring the consistency of the values used is a basic requirement for high-quality analysis of financial results.

Cash Return and Cash Flow

Cash flow is at least as important as profits. This concept refers to how much cash is available to an organization to (a) pay current bills on time, (b) take advantage of discounts or volume purchases, (c) build adequate inventory levels, (d) carry the value of accounts receivable, and (e) pay for expansion of markets or products.

Using cash flow as a measurement of investment success, you can measure how long it should take to break even. The question, "How long will it take to get back my investment?" (based on cash flow) is addressed by the payback ratio. This issue comes up whenever a business places cash into a new product, for example, or expands geographically. Management needs to know how much commitment is involved. The commitment for expansion is not limited to the cash, but also includes the time needed to recapture that cash.

The payback ratio compares the initial investment to net cash flow. The term "net cash flow" means the total of revenues less costs, expenses, and taxes, all expressed on a cash basis. (Non-cash items such as depreciation or accruals are left out of net cash flow.) The formula:

Formula: Payback Ratio

$$I \div C = R$$

where: I = initial cash investment
C = net cash flow per year
R = payback ratio

On a spreadsheet program, enter the following values:

A1   I
B1   C
C1   =SUM(A1/B1)

Example: Your company has invested $235,000 to expand its geographic market over the next year. The question has been raised, "How long will it take to recapture this investment based on increased profits?" The analysis of this plan originally was prepared in the usual forecasting and budgeting model and did not adjust to cash-only. A revised forecast estimated that net cash flow per year will average $41,500. Enter these fields on the spreadsheet:

A1   235,000
B1   41,500
C1   =SUM(A1/B1)

The payback ratio is:

$$\$235,000 \div \$41,500 = 5.7 \, years$$

The forecast indicates that it will take 5.7 years to break even on this expansion based on the forecast model being used. For management, the ultimate decision to proceed or not should rely on whether the forecast is reliable, and on whether an investment of $235,000 for nearly six years is prudent. Even if long-term profits will grow beyond the forecast level, management must be concerned with the level and time of the commitment. This makes the payback ratio a valuable test of long-term return on investment.

A closely related test of return is called cash-on-cash return (also called equity dividend yield). This is a test of the actual return based on cash flow per year and the formula is the opposite expression of the payback ratio:

Formula: Cash-on-Cash Return

$$C \div I = R$$

> where: C = net cash flow per year
> I = initial cash investment
> R = cash-on-cash return

In a spreadsheet program, enter the following values:
    A1   C
    B1   I
    C1   =SUM(A1/B1)

Using the previous example, the cash-on-cash return is expressed as a percentage:

$$\$41,500 \div \$235,000 = 17.7\%$$

Enter these values:
    A1   41,500
    B1   235,000
    C1   =SUM(A1/B1)

Although this is simply the reverse calculation of the payback ratio, it is also revealing. Management knows it will take 5.7 years to recapture its initial investment based on forecast annual cash flow. This is instructive when making judgments about the level of cash and the time required to recapture it. Cash-on-cash adds to the information by reducing the question to a cash-based rate of return on the investment. This is especially useful when comparing the planned expansion proposal to current operations. If the current configuration is earning net return of 9.0% per year, the expansion program is going to be more aggressive than current income levels. If the cash-on-cash return comes out lower than current net profits, the question must be raised: "Is this expansion proposal a good idea?"

Cash flow and net income are worth comparing at times of planned expansion, as well as for valuation of equity itself. Cash flow's impact has three key elements:

First, it is important ... to understand the difference between net income and cash flow and to recognize that cash flow trumps net income in assessing a firm's financial performance or, as some finance textbooks state, "cash is king." Second, understanding the importance of a firm's cash flow sets the stage for later discussing the valuation of bonds, stocks, capital investment projects, and even the firm itself, all of which involve discounted cash flow analysis. Finally, a firm's cash flow ... has implications for the firm's owners and its managers ...[2]

Developing and comparing cash flow-based returns, management can determine (1) the time required to recapture the cash outlay, (2) cash-based return based on the forecast, and (3) profitability of the expansion compared to current profitability of the organization.

In studying and comparing the cash-on-cash return to current net profits, it is not entirely accurate to use the tax-based profits of the organization. Since this includes non-cash expenses, it is not comparable to the cash-on-cash return. To make the two sides of the equation comparable, current income should be adjusted to the cash basis. This does not require completely revamping reporting profits; for comparisons between the organization's profits and cash-on-cash return, the only recurring non-cash expense, depreciation, should be added back in to the reported profits. (Depreciation calculations are explained in Chapter 7). The formula for cash income:

Formula: Cash Income

$$I + D = C$$

where: I = net income
D = depreciation expense
C = cash income

On a spreadsheet program, enter the following values:
A1 I
B1 D
C1 =SUM(A1+B1)

Enter these values:
A1 1,450,000
B1 850,000
C1 =SUM(A1+B1)

---

2 Petty, J., & Rose, J. (2009). Free Cash Flow, the Cash Flow Identity, And the Accounting Statement of Cash Flows. *Journal of Financial Education*, 35, 41–55.

Example: You will recall that last year's net income was $1,450,600 against revenues of $16,118,000. Net return was:

$$\$1,450,600 \div \$16,118,000 = 9.0\%$$

The initial comparison between this net return and a planned expansion generating 17.7% cash-on-cash return seems to point to an obvious conclusion: The expansion makes perfect sense. However, annual depreciation expense is approximately $850,000. Applying the cash income formula:

$$\$1,450,600 + \$850,000 = \$2,300,600$$

This changes the picture. Now the net return using cash income is:

$$\$2,300,600 \div \$16,118,000 = 14.3\%$$

Although this adjusted calculation is still below the forecast 17.7% cash-on-cash return expected from expansion, the cash income is quite close. It raises a new question: Is the expansion feasible given its proximity to current cash income? There is some doubt. The comparison is being made between known historical income and a forecast outcome. If the forecast is off over the next 5.7 years and less additional cash profit is generated, the company might not recapture its investment for many years beyond.

   This does not indicate that the expansion is an ill-conceived idea. It does point to the possibility that the risk of that much cash, outstanding for so many years, may be too high. The judgment call is not merely a comparison between returns of 9.0% and 17.7%. The estimated outcomes are much closer when analyzed on a cash basis. Management makes better decisions when its information is reliable. The preceding example demonstrates that when current profitability and forecast expansion profitability are both compared on a net cash basis, the likely outcomes are much closer than the original numbers indicated.

Returns on Purchases and Sales

Every organization invests in money in capital assets or marketable securities. Capital assets—equipment, machinery, vehicles, and real estate—are depreciated over a recovery period and the depreciation is deductible from revenues as part of the income statement. When expenses are considered as part of the return calculation, a few variations of "return" can result. These include the return on purchase price, return on cash investment, and return on net investment. When you are dealing with the return on an investment that includes income in the form of either interest or dividends, the "total return" must include these as part of the equation.

The return on purchase price is a calculation of the return on the entire price, regardless of whether it was partially financed, and without considering depreciation and other costs. This calculation also does not consider the holding period. To calculate, divide the net difference between sale and purchase, by the original purchase price. The formula:

Formula: Return on Purchase Price

$$(S - P) \div P = R$$

where: S = sales price
P = purchase price
R = return on purchase price

Example: You purchased a capital asset for $16,500 two years ago. This year you sold it for $19,000. Return on purchase price is:

$$(\$19{,}000 - \$16{,}500) \div \$16{,}500 = 15.2\%$$

On a spreadsheet program, enter:
    A1   S
    B1   P
    C1   =SUM(A1-B1)/B1

When you purchase an asset for $19,000 but you finance $17,000 of the price, the return on purchase price does not tell the entire story. Obviously, the outcome is different if you pay all cash versus borrowing most of the price. A more reliable calculation is return on cash invested (also called return on invested capital). The formula:

Formula: Return on Cash Invested

$$(S - P) \div I = R$$

where: S = sales price
P = purchase price
I = cash invested
R = return on cash invested

Example: In the previous example, you bought the asset for $19,000 but financed $14,000; you invested $5,500. When that asset was sold for $19,500, the return on cash invested was much different than the previously calculated return on purchase price. However, this calculation does not consider the cost of borrowed money, depreciation, or holding period. Applying the formula:

$$($19,000 - $16,500) \div $5,500 = 45.5\%$$

This outcome measures only the net profit versus cash invested and, because it does not take the entire cash difference or time into account, is not a reliable measurement of profitability on the transaction. However, it is useful in judging the effectiveness of borrowing money to invest. For example, if the transaction had resulted in a loss, then the formula could be used to calculate the negative effects of borrowing money, which becomes useful when a series of similar transactions are compared.

This formula can also be summarized on a spreadsheet. Enter the following fields:

A1 S
B1 P
C1 I
D1 =SUM(A1-B1)/C1

Applying the example, entries are:

A1 19000
B1 16500
C1 5500

The next formula in the progression requires expressing the outcome on a realistic basis. This requires taking costs into account. The "net investment" is the actual cash put into an investment, adjusted downward for the costs involved. If you borrow money to finance the purchase, the return must be reduced by the interest expense related to the loan. The formula for return on net investment does that, making the calculation more valuable in comparing two or more outcomes. The formula:

Formula: Return on Net Investment

$$(S - P - C) \div I = R$$

where: S = sales price
       P = purchase price
       C = costs
       I = cash invested

R = return on cash invested

Example: If you bought an asset for $16,500 with a $5,500 investment, the return on cash invested when selling for $19,000 (previously calculated) provides part of the answer but does not consider the cost of borrowing money. If interest came up to $1,700 in this case, the calculation would have to reduce net return by that amount:

$$(\$19,000 - \$16,500 - \$1,700) \div \$5,500 = 14.5\%$$

This is considerably lower than the previously calculated 45.5%, which excluded interest. However, the cost of borrowing money is a substantial factor in calculation of overall return. This formula can also be reflected in spreadsheet entries:

    A1 S
    B1 P
    C1 C
    D1 I
    E1 =SUM(A1-B1-C1)/D1

Applying the example, the entries are:

    A1 19000
    B1 16500
    C1 1700
    D1 5500

The inclusion of costs is critical to arriving at an accurate formula for the net cash return on an investment. Costs may also include a variety of items beyond interest needed to arrive at the true net difference. In some kinds of investments, the adjustment needs to also include the effects of earnings. For example, if you buy shares of a mutual fund with temporarily excess cash, the net return should be adjusted upward for any interest, dividends or capital gains received during the holding period. The calculation is the same as that to account for costs; however, the added income would be an addition to the cash-based profit rather than a reduction.

Investment-based Returns

In addition to calculating return on revenue or capital, you may also need to calculate return on investments, in various forms. These calculations cover a range of topics, including dividend yield, current yield on bonds, discount yield (on real estate) and earnings expressed on a per-share basis.

Dividend yield (also called "current yield" by stock analysts) is the percentage represented when dividend per share is divided by the current price per share.

Divided yield is more significant than the percentage or dollar value received by owning stock. The level of dividend yield often serves as a basis for deciding to invest in one company over another:

Dividend-yield strategies belong to the broader class of value investment strategies. Value stocks are characterized by high dividend yields ... Growth stocks are classified as stocks exhibiting the opposite characteristics ... numerous studies have found that long term value strategies tend to outperform the market in the U.S. financial markets.[3]

The formula:
Formula: Dividend Yield

$$D \div P = Y$$

where: D = dividend
P = price per share
Y = dividend yield

Example: You are considering investing in stock current selling for $52 per share. The company's announced dividend per share is $1.40. Dividend per share is:

$$\$1.40 \div \$52 = 2.7\%$$

The formula can be summarized on a spreadsheet program, using the following values:

A1   D
B1   P
C1   =SUM(A1/B1)

In calculating dividend yield, remember that the outcome is going to vary every time the price per share changes. Accordingly, this calculation applies in two circumstances. First, when considering investing in shares, the yield applies to the current price. Second, after purchase, the yield you calculated at the time of purchase is the applicable yield, even if the price has changed significantly. The higher the price, the lower the yield for a dividend that has not changed. In the previous example, stock was selling for $52 per share and dividend yield was 2.7% based on an annual dividend of $1.40 per share. However, if the stock price rose to $55 per share, yield would fall to 2.5% ($1.40 ÷ $55); and if the price per share fell to $50, yield would rise to 2.8%

3 Visscher, S. & Filbeck, G. (2003). Dividend-Yield Strategies in the Canadian Stock Market. *Financial Analysts Journal*, 59(1), 99–106.

($1.40 ÷ $50). The yield an investor earns is based on the price paid per share, and not on the ever-changing price per share.

Another form of current yield applies to bonds. This is a more complex calculation because it is based on the changing current value of the bond. A bond may sell at par (100) or at a discount or premium below or above par. This is caused by the changing market interest rate compared to the fixed rate paid by the bond. When bond interest is higher than market rates, the bond is likely to sell at a premium. To calculate the current yield, the nominal (stated) yield is divided by the current value (premium or discount). This produces the current yield. A bond selling at par will reflect a current yield equal to its nominal yield. However, whenever the current value is above or below, current yield will change as well. The formula:

Formula: Current Yield on a Bond

$$N \div V = C$$

> where: N = nominal yield
> V = current value of the bond
> C = current yield

Example: The 3% bond you purchased for $1,000 is current selling at a discount of $980. In bond jargon, the bond in this condition is described as selling at 98. Current yield is:

$$3.00 \div 98 = 3.1\%$$

If the bond were later to appreciate to 103 (a premium 3% above par, or $1,030), the formula would be:

$$3.00 \div 103 = 2.9\%$$

The formula can also be summarized on a spreadsheet, using the following cell values:

A1   N
B1   V
C1   =SUM(A1/B1)

A popular investment calculation is earnings per share (EPS), which is simply the company's earnings expressed on a per-share basis. To calculate, the latest known earnings per year are divided by the number of shares outstanding. The formula:

Formula: Earnings Per Share (EPS)

$$E \div S = EPS$$

where: E = earnings
S = number of common shares outstanding
EPS = earnings per share

Example: A corporation currently has 1,400,000 shares outstanding. The recently released annual earnings report showed a total of $982,000. EPS is:

$$\$982,000 \div 1,400,000 = \$0.70$$

The outcome, or 70 cents per share, is valuable when a long-term EPS trend is studied. Most financial ratios are most revealing when you can see whether earnings are improving or declining over many years. The spreadsheet cell values for this formula are:

A1   E
B1   S
C1   =SUM(A1/B1)

The EPS is also used in a related formula that is widely used by investors, the price/earnings ratio, or PE. This is the price per share divided by earnings per share. The formula:

Formula: Price/Earnings Ratio

$$P \div E = PE$$

where: P = price per share
E = EPS
PE = price/earnings ratio

Example: The price per share of stock in a company you are tracking is $35. The most recent EPS was $1.75 (earnings represented $1.75 per share). The PE was:

$$\$35 \div \$1.75 = 20$$

To interpret the meaning of the PE, the result, also called the "multiple," is the number of years of earnings represented by the current price per share. In this example, the multiple of 20 is equal to 20 years of earnings for the company. The higher the PE, the greater the market risk. While opinions vary, it is generally understood that PE of 25 or below is considered a relatively safe investment, and PE above 25 is increasingly inflated in value.

The PE can also be summarized on a spreadsheet using the following cell values:

A1   P
B1   E
C1   =SUM(A1/B1)

PE is widely used as a means for comparing companies to one another or for tracking a company's market risk over many years. It is an especially valuable ratio because it combines a technical indicator (price) with a fundamental indicator (earnings). One potential problem with this is a disparity of timing. If the earnings are based on information available three months ago, comparing the reported outcome to the current price could be unreliable or distorted. For this reason, the *forward* PE is often used in place of the more traditional ratio based on potentially outdated earnings. The forward PE is based on currently estimated or expected earnings per share. The formula is identical to that for PE; however, expected EPS is used in place of the latest reported earnings.

Annualized Return

All return calculations are based on calculations between profits and some form of basis (purchase price, equity, or sales, for example). However, the reported return must vary based on the period involved. A 4% return over three months is much higher than a 4% return over 15 months. Whenever return applies for a period other than exactly one full year, the rate earned must be annualized. This means it is expressed as if the period in effect was exactly 12 months. For any comparisons between dissimilar holding periods, annualization is essential to ensure accuracy.

To annualize a return, divide the percentage of return by the holding period expressed in months, and then multiply the result by 12. The use of the number of months is reliable for most applications; however, you can also annualize using weeks or an exact number of days. The formula to annualize a reported return:

Formula: Annualized Return

$$(R \div H) \times Y = A$$

where: R = return
      H = holding period
      Y = periods in one year (12)
      A = annualized return

Example: You have invested funds in two different stocks as part of your business money management program. The stocks both earn 4%. However, one is held for seven months and the other for 15 months. The annualized rate is not the same in this case. If the annualized return is going to be based on the months each investment was owned, the valued of 'Y' is 12, representing the number of months in one full year. To annualize each of the returns:

7-month holding period: $(4.0 \div 7) \times 12 = 6.9\%$
15-month holding period: $(4.0 \div 15) \times 12 = 3.2\%$

The shorter the holding period, the higher the annualized return, and vice versa. The purpose in annualizing is to ensure that comparisons are valid. In the example, the two 4% returns are not equal because the holding period was dissimilar. The 7-month return (annualized at 6.9%) was more than twice as profitable as the 15-month return (annualized at 3.2%). The formula can also be reduced to a spreadsheet calculation (based on the assumed use of 12 months):

A1 R
B1 H
C1 =SUM(A1/B1)*12

Calculating rates of return is not a difficult or complex process. The complexity comes in defining exactly which values to use in the formulas. Annualizing is also crucial to making valid comparisons between two or more outcomes. In the next chapter, the same basic assumptions and qualifications are applied to the calculation of a topic every manager eventually must contend with: cash flow and leverage.

Chapter 4
Calculating Breakeven and After-Tax Profit

How do you know whether you are profiting from the use of money? A manager who does not accurately summarize profit and loss can easily fall into a reporting trap: If profits have been earned when, in fact, a net loss has taken place.

This occurs when you overlook the effects of hidden costs, inflation, and taxes. These all affect your estimate of profits; when considered collectively, they can be significant. To accurately estimate profits, you need to develop reliable methods for estimating hidden costs, inflation and tax consequences. These are part of any financial summary or cash flow analysis. The profitability of any activity relies on the accuracy of these estimates.

Hidden Costs

Because hidden costs cannot be precisely known in advance, some methodical estimates must be used to take them into account. The procedure for many unknown hidden costs is to set up a loss reserve. This reserve is carried as an expense during each period, offset by a reduction in an asset's book value. For example, a bad debt reserve is established to record bad debts on a regular basis even though these will occur in the future. The purpose of the reserve is to record expense in each reporting period rather than all at once.

Like so many estimates entered in a company's books, a reserve for bad debts can be manipulated and misused to meet or beat forecasts or create artificial adjustments. For example, in good years, reserves can be overstated so that future periods with disappointing results can be improved—a method called "cookie jar" accounting. One study concluded that:

> Firms manage bad debt expense downward (and even record *income-increasing* bad debt expense) to meet or beat analysts' earnings forecasts … firms manage bad debt expense downward by drawing down previously recorded over-accruals of bad debt expense that have accumulated on the balance sheet.[1]

The implications of these practices include concern that accounting results are not always entirely accurate. A good deal of interpretation (and even manipulation) might be involved. For this reason, a critical analysis of accounting decisions is wise, if only to ensure that yearly trends remain reasonable and are not altered simply to

---

1 Jackson S., & Xiaotao, L. (2010). The Allowance for Uncollectible Accounts, Conservatism, and Earnings Management. *Journal of Accounting Research*, 48(3), 565–601.

DOI 10.1515/9781547400638-004

maintain expectations. The reserve method is probably a leading source of question-able accounting decisions.

The accounting theory behind the reserve method is that expenses occur over time and matters like writing off bad debts should not be reported only in the year when a company realizes a debt has become uncollectible. The reserve is created to "recognize" a portion of future bad debts each year. The principle of recognition is a reference to the accounting period where an expense is recorded. An observer outside of the accounting department might easily assume that an objective standard has been applied in identifying the level of reserve. This is not always the case.

A bad debt is recorded by reducing the current asset "Accounts Receivable" by the reserve amount, offset by a current-year expense. For example, if a company writes off $1,500 per quarter for bad debts, the quarterly entry is:

	Debit	**Credit**
Bad Debt Expense	$1,500.00	
Reserve for Bad Debts		$1,500.00

In the asset section of the balance sheet, the bad debt reserve reduces the value of the Accounts Receivable account. If a company has been putting aside a reserve of $1,500 per quarter for the past five quarters, and current Accounts Receivable are $455,000, the status of these two accounts is:

Accounts Receivable	$455,000
Less: Reserve for Bad Debts	– 7,500
Net Accounts Receivable	$447,500

The company has recognized $1,500 each quarter rather than having the future bad debts expense hit the books all at one time. This makes the reporting more orderly and consistent and is an appropriate application of the reserve method. For example, if the company writes off a bad debt of $6,200, the write-off goes against the reserve and not against the expense that has been accumulating during the preceding five quarters. The entry to record a bad debt reduces both sides of the asset account:

	Debit	**Credit**
Reserve for Bad Debts	$6,000.00	
Accounts Receivable		$6,000.00

After this entry has been made, the two balance sheet accounts reflect the change:

Accounts Receivable	$449,000
Less: Reserve for Bad Debts	– 1,500
Net Accounts Receivable	$447,500

By using this system, each quarter shows a bad debt expense of $1,500.00 and the reserve accumulates until accounts are written off. The level of bad debts is determined by historical levels of bad debts. If a company experiences 1.5% of bad debts on average, the reserve should be set at that level to anticipate that the trend will continue. As actual experience comes about higher or lower than the historical level, the reserve should be adjusted as well.

This summarized version of how hidden costs are estimated and placed into the books is only one way that they can be managed. No one likes surprises and managers are no exception. You would not want to have to explain why the current quarter's profits were unexpectedly reduced by $6,000 because no bad debt reserve had been established.

A similar approach can be used to set up reserves for any number of expenses. In companies with a lot of high-maintenance machinery, a deferred maintenance reserve can be set up to anticipate expensive periodic repairs. If machinery typically needs to be overhauled every four years, it makes sense to set up a current reserve to book the known future expense every quarter:

> All machinery is on an irresistible march to the junk heap, and its progress, while it may be delayed, cannot be prevented by repairs ... in valuing all fixed assets, an account must be taken of the lapse of time, and even in the case of machinery giving no evidence either of use or misuse, the bare fact that it is a year nearer its inevitable goal is an item of which technical account must be taken.[2]

In cases when relatively new equipment requires little or no maintenance, the level of deferred maintenance expense may be used to justify establishing a reserve for deferred maintenance. This is separate from the expected depreciation write-off each year. Whereas depreciation recognizes annual levels of the original cost, maintenance is separate and above capital investment.

Equipment may also become obsolete as new models are developed and placed on the market. However, an obsolescence reserve is not normally required since this is assumed to be a part of the depreciation system. Capital assets are booked onto the balance sheet and then depreciated over a specified "recovery period." This is a reference to the write-off of the cost of buying and holding capital assets. Over the recovery period, the value of the asset is gradually set up as an expense and the purchase basis of the asset is reduced. (See Chapter 7 for more information concerning depreciation.)

A depreciation system serves two purposes. It sets up a reserve for possible obsolescence while also providing the method for converting the asset to expense over several accounting periods. The rationale for this is the same as that for setting up a bad debt reserve. It would distort this year's operating results to write off the entire cost of buying new equipment as an expense. By definition capital assets will last for

---

2 Hatfield, H. R. (1915). *Accounting: Its Principles and Problems*. New York: D. Appleton, p. 121.

several years; they are depreciated over a period estimated to represent the "useful life" of the asset.

This is not always realistic. For example, in the case of real estate, a purchase must be placed on the books at actual purchase price and then depreciated over many years. Two factors distort this procedure. First, land cannot be depreciated, so its value remains unchanged for the entire time your company owns it. Second, the book value of real estate declines each year as depreciation is recorded, even when the market value of real estate rises. The accounting convention does not allow you to adjust the net book value (purchase price minus depreciation) to reflect its current value, so these adjustments must be reported in footnotes. The net book value does not always reflect the reality. Real estate may be worth much more, and some other assets (notably vehicles) might be worth much less. It all depends on how value changes in the market, compared to a pre-established depreciation schedule.

Example: Your company has two assets on the books that are valued improperly. Many years ago, the company purchased its headquarters building for $2.6 million. Today, the book value is approximately $1.9 million net of depreciation. However, the company recently refinanced its mortgage and the latest appraisal concluded that the property is now worth $4 million. The second asset is a truck purchased for $28,000. Net book value today is $16,000 but because the newest models provide greater fuel efficiency and lower-maintenance engines, the true market value of the truck is esti-mated at only $2,000. The adjustment between net book value and market value are:

	Net book value	Market value	Difference
Real estate	$1,900,000	$4,000,000	($2,100,000)
Vehicle	16,000	2,000	14,000

The vehicle's disparity is quite small compared to the real estate change between net book value and market value. This is a common problem, caused by the accounting restriction and the unyielding rule that an asset's acquisition value, minus deprecia-tion, must be the reported book value of that asset. These kinds of disparities in how assets are valued can lead to problems for companies. In the 1980s when leveraged buyouts were popular, some companies were taken over for a price above book value but below their true market value. If management is not aware of the market value of its capital assets, then the entire financial reporting system can deceive rather than enlighten any analyst.

Hidden costs can involve many other sections of a company's financial report beyond such apparent and visible ones like bad debt reserves or accumulated depre-ciation. A reserve to cover uninsured losses might be needed in a company vulner-able to litigation but lacking complete insurance coverage. The interest expense of long-term financing without a contractual fixed interest rate poses another form of potential future loss. Inadequate cash flow represents a lost opportunity risk because the company will not be able to act if expansion opportunities or mergers and acquisitions become possible. Without cash, many opportunities must be passed up without acting. So hidden costs come in many forms and a realistic approach to long-term planning requires identification of those costs and when possible, either setting up reserves or changing the current operating model to reduce or eliminate the problem.

Some small companies, for example, have eliminated the expense of chronic bad debts by simply refusing to extend credit to customers. They require all bills to be paid at the time goods or services are delivered. This move is not always practical, and invariably leads to the loss of some business; however, if bad debts have impacted profits to the extent that it is no longer affordable, the loss of business becomes more acceptable than growing losses each year. The combined negative impact on profits and cash flow makes decisions like this unavoidable.

The Inflation Effect

Hidden costs are likely to destroy even the best thought-out long-term forecast and replace profits with losses. Even if your company sets aside reserves for hidden costs, they have an eroding effect. One of the most hidden and most destructive of these is the impact of inflation.

The meaning of inflation is misunderstood by many people. Some believe that it is caused by rising prices when, in fact, higher prices are a *symptom* of inflation and not its cause. The confusion was explained many years ago by Roger Blough, CEO of U.S. Steel, who observed that "Steel prices cause inflation like wet sidewalks cause rain."[3]

The real definition of inflation is the gradual erosion of purchasing power. As the dollar's value falls, it takes more to buy the same number of goods or services. For example, one dollar in 1950 was the equivalent of $7.15 in the year 2000.[4]

Valuable Resource: The Bureau of Labor Statistics provides a free inflation calculator. You can enter any two years to determine how purchasing power changes over time. Go to www.bls.gov/data/inflation_calculator.htm to use the calculator.

This means that you needed $7.15 in 2000 to buy $1.00 worth of goods based on 1950. The difference, or the loss of purchasing power, comes from inflation over 50 years. The annual or quarterly inflation rate is measured by the Consumer Price Index (CPI), published by the Bureau of Labor Statistics. This is a study of a basket of consumer goods, calculated against an index level with a percentage of change each year. Infla-

---

3 Blough, Roger (CEO of U.S. Steel). (August 1, 1967) *Forbes.*
4 Bureau of Labor Statistics (www.bls.gov).

tion is caused primarily by changes in the money supply. As more money is printed and distributed, its real value declines. Whenever the money supply grows faster than economic growth (normally measured by Gross Domestic Product, or GNP), inflation rises. Excessive levels of money are printed to fund growing government debt or to improve the credit markets; but over the long term, the effects of inflation create the unavoidable deterioration in purchasing power. For managers, inflation is important not only because it affects purchasing power on both a personal and corporate level, but also because it directly affects all calculations of future cash flow and profits.

The rate of inflation, which is expressed as a percentage of increase in a specified period (month or year, for example) is based on the use of an index developed by the Bureau of Labor Statistics (BLS). To calculate annual inflation, subtract the change in the annual index and divide the difference by the earlier year's index level. The formula:

Formula: Inflation Rate

$$(C - P) \div P = I$$

      where: C = current CPI index
              P = past CPI index
              I = rate of inflation, CPI

Example: The Bureau of Labor Statistics reported that its index level at the end of 2008 was 210.2. At the end of 2007, it was 210.0 and at the end of 2001 it was 176.7. These values are based on an index of "100" in 1983. Applying the formula to the seven-year change between the end of 2001 and 2008 (using the CPI for "All Urban Consumers, also called CPI-U):

$$(210.2 - 176.7) \div 176.7 = 19.0\%$$

The formula can be summarized on a spreadsheet, using the following field values:

    A1   C
    B1   P
    C1   =SUM(A1-B1)/B1

From a business manager's perspective, the consequences of inflation are likely to be far different than the consequences for an individual consumer. This difference comes from the way that one set of inflated prices affects overall costs to produce, market, and sell goods. For example, the consumer is going to suffer from inflated prices of gasoline and may be less aware that gas prices also affect the cost of food, utilities, and even housing. A business manager is likely to be more aware that the overall, inflation rate is not equally distributed. As a result, some costs are not as obvious as others. For example, a lender or savings institution with existing contracts on the books for fixed interest income or expense is going to be quite sensitive to changes in interest rates. An organization providing non-essential products or services (such as a recreational company, travel or tourism outlet, or hotel chain, for example) will be far more sensitive to the immediate effect of inflation than a company selling necessities. This problem is especially aggravated in times of hyper-inflation, when the rate is either out of control or rising rapidly.

In calculating a breakeven point, inflation must be brought into the equation. For managers, the concept of breakeven is the identification of the minimum volume of revenues produced to justify an activity. If this forecasting device does not include a factor for inflation, it will be unrealistic, especially in cases when long-term projections are being offered.

Example: Your company is evaluating a multimillion-dollar investment in a new product line and initial estimates indicate that if successful, breakeven will occur within five years. The forecasts include a fixed price and fixed cost assumption and are based on current expense levels. The breakeven sales level on an annual basis is:

Revenues	$14,400,300
Less: Direct Costs, 58%	– 8,352,174
Gross Profit	$ 6,048,126
Expenses (fixed assumption)	– 6,000,000
Net Pretax Profit	$ 48,126 (0.3%)

This set of assumptions includes a belief that the expense level and the percentage of gross profit will remain unchanged. The product becomes profitable if revenue levels rise above this breakeven. For example, if revenues were to rise to $30 million per year:

Revenues	$30,000,000	
Less: Direct Costs, 58%	– 17,400,000	
Gross Profit	$12,600,000	
Expenses (fixed assumption)	– 6,000,000	
Net Pretax Profit	$ 6,600,000	(22.0%)

These assumptions could be accurate assuming that no inflation occurs. However, what if the cost of goods sold were to rise? What if expense levels increased substantially? One of the assumptions used was that the product's price would be fixed to remain competitive with better-established brands. The estimated 22% return is not going to be accurate if the revenues remain unchanged, but direct costs were to rise by 8% over 10 years and expenses were to rise by 15%:

Revenues	$30,000,000
Less: Direct Costs, 68%	– 20,400,000

Gross Profit	$ 9,600,000	
Expenses (fixed assumption)	− 6,900,000	
Net Pretax Profit	$ 2,700,000	(9.0%)

This forecast is less satisfying that the previous assumption of 22%, but it might also be more realistic. Given the altered numbers with inflation assumptions included, this type of exercise should be built into any long-term revenue and profit forecast.

Taxes in the Profitability Equation

Besides inflation, the tax reality should be included in forecasts as well, and not only in long-range ones but even those looking out only one year. Remember:
– Taxes apply to *all* net income from operations.
– In addition to federal income taxes, state tax rates must be included.
– A forecast is incomplete if it excludes the tax factor in the outcome.

The corporate tax rates apply both federally and in most states. (Nine states—Nevada, South Dakota, Texas, Washington, Alaska, Wyoming, Florida, Tennessee, and New Hampshire—assess no income tax). As of 2018, the federal rates were:[5]

Rate	Individuals	Married Filing Jointly
10%	Up to $9,525	Up to $19,050
12%	$9,526 to $38,700	$19,051 to $77,400
22%	$38,701 to $82,500	$77,401 to 165,000
24%	$82,501 to $157,500	$165,001 to $315,000
32%	$157,501 to $200,000	$315,001 to $400,000
35%	$200,001 to $500,000	$400,001 to $600,000
37%	over $500,000	over $600,000

State tax rates vary by state. Some examples of top marginal rates: [6]

California	13.3%
Oregon	9.9%
Massachusetts	9.5%
New Jersey	8.97%
Pennsylvania	3.07%
North Dakota	2.9%

---

5 Berger, Ron (December 17, 2017). The New 2018 Federal Income Tax Brackets & Rates. *Forbes.*
6 https://taxfoundation.org

Valuable Resource: To review the state tax rules and conditions in your state, go to the website https://taxfoundation.org

---

If your company is based in California and your taxable income is $12 million, your combined taxes will be:

Federal 37%	$4,440,000	
State 13.3%	1,596,000	
Total tax	$6,036,000	(50.30%)

The high combined federal and state corporate tax rates create a very different picture than you get from the pre-tax profit. The after-tax return is the pre-tax return minus the effective combined tax rate, and this is the number you must use to estimate a realistic income level. The calculation:

Formula: After-tax Return

$$O \times (1 - T) = A$$

where: O = operating (pre-tax) profit

T = combined federal and state tax rate (in decimal form)

A = after-tax profit

Example: Your company's pre-tax profit is $12 million. Your federal rate is 37% and your state tax rate is 13.3%. The formula to determine after-tax profit is:

$$\$12,000,000 \times (1 - .503) = \$5,964,000$$

If this income represented the operating profit on $150 million, the pre-tax level would be 8% ($12,000,000 ÷ $150,000,000). However, the after-tax return represents only 3.98% ($5,964,000 ÷ $150,000,000). In evaluating and comparing net return, the use of the after-tax formula makes more sense, especially in states with high tax rates.

The after-tax return can be calculated on a spreadsheet as well:
A1   O
B1   T (in decimal form)
C1   =SUM(1-B1)*A1

Based on the previous example, this produces the values:
A1   12,000,000
B1   .503

C1 $5,964,000

The importance of calculating federal and state taxes as part of the analysis and fore-casting of revenues and profits cannot be emphasized too much. Taxes are a key factor and play as important a role as that of inflation. However, when you consider the effect of inflation and taxes together in any estimate of profitability, the entire picture is likely to change even more drastically than when one of these is considered alone.

Breakeven Calculations: Inflation *and* Taxes

As serious an impact as inflation and taxes have on true profitability, when they are combined, their affect is substantial. This reality is overlooked in most of studies, whether aimed at long-term financial results or the shorter-term investment returns and cash flow within a single business cycle. As a result, projections lacking consid-eration of the tax and inflation impact are unrealistic.

Why do these have to be considered together? The impact of inflation and taxes on profitability is unavoidable, even though it is not always taken into consideration. The argument could be made that if comparisons are made on a like-kind basis, this does not matter. In other words, if the inflation and tax factors are excluded in each case, the comparison is still valid. This is not always the case, however, for several reasons:

1.   Taxes vary by state where most businesses are based; if two companies under review are active in different jurisdictions, the tax impact of each will be quite different.
2.   Federal taxes change based on dollar levels of taxable profits. As a result, com-parisons between companies with dissimilar net profit are invalidated if the tax impact is excluded. This becomes most important when projecting future growth in profitability. Under a current level of profits, taxes might be insignificant, but given assumptions about (a) future growth and (b) the resulting changes in effec-tive tax rates, optimistic forecasts of revenues and profits that exclude consider-ation of federal taxes make projections increasingly unrealistic.
3.   Inflation does not impact every organization in the same way. A study of the ele-ments making up the CPI show that the collective number is the weighted average of many different products and in many different areas. So different companies are going to be subject to different net inflation rates. For example, a company that relies on transportation of goods will be more impacted by higher oil and gas inflation, than one providing a service sold over the Internet.
4.   Although general assumptions and estimates of inflation can be applied univer-sally, the specific sector must be considered to determine how much weight to give to inflation. If two or more organizations are going to suffer identical infla-tion is valid only if they are active in the same primary sector.

5. When you calculate net return based on adjustments for inflation and taxes, the true result makes more sense. This assumes that a detailed analysis has been made to adjust for varying tax rates by location and income levels, and that the assumptions about inflation take specific factors into account. Thus, a published annual CPI may be weighted higher or lower based on how the elements of CPI are calculated. (Check the Bureau of Labor Statistics website for a detailed explanation of the components of CPI: www.bls.gov.)

You need to know your net breakeven return to determine whether profits are truly "profitable." If in fact you are losing net purchasing power after inflation and taxes, then there is a problem. The calculation of breakeven return is based on the combined effects of inflation and taxes. To calculate:

Formula: Breakeven Return

$$I \div (1 - T) = B$$

> where: I = inflation rate in decimal form
> T = effective tax rate, including both federal and state (in decimal form)
> B = breakeven return

Example: Your company invests temporarily available cash in a carefully selected portfolio of money market funds and high-dividend stocks. The portfolio return has averaged 4.5% over the past year, which management considers quite positive. But is it above breakeven? Based on an assumed 3% annual inflation and combined federal and state taxes at 43.84%, breakeven return in this example is:

$$3.0\% \div (1 - .4384) = 5.3\%$$

Given these assumptions, you need to earn 5.3% on the portfolio just to break even. If you earn anything less than this, you are losing money on a net-of-inflation and net-of-tax basis. If you duplicate the 5.3% rate, you are simply holding purchasing power at current levels. Reacting to the 4.5% average return as a *profit* is not realistic.

The formula can be expressed on a spreadsheet, using the following values:
A1   I (in decimal form)
B1   T (in decimal form)
C1   =SUM(A1/(1-B1))

The formula is also summarized by various rates of inflation and tax levels, in Table 4.1.

Table 4.1: Breakeven return

Effective tax rate	INFLATION RATE					
	1%	2%	3%	4%	5%	6%
14%	1.2%	2.3%	3.5%	4.7%	5.8%	7.0%
16	1.2	2.4	3.6	4.8	6.0	7.1
18	1.2	2.4	3.7	4.9	6.1	7.3
20	1.3	2.5	3.8	5.0	6.3	7.5
22	1.3	2.6	3.8	5.1	6.4	7.7
24%	1.3%	2.6%	3.9%	5.3%	6.6%	7.9%
26	1.4	2.7	4.1	5.4	6.8	8.1
28	1.4	2.8	4.2	5.6	6.9	8.3
30	1.4	2.9	4.3	5.7	7.1	8.6
32	1.5	2.9	4.4	5.9	7.4	8.8
34%	1.5%	3.0%	4.5%	6.1%	7.6%	9.1
36	1.6	3.1	4.7	6.3	7.8	9.4
38	1.6	3.2	4.8	6.5	8.1	9.7
40	1.7	3.3	5.0	6.7	8.3	10.0
42	1.7	3.4	5.2	6.9	8.6	10.3
44%	1.8%	3.6%	5.4%	7.1%	8.9%	10.7%
46	1.9	3.7	5.6	7.4	9.3	11.1
48	1.9	3.8	5.8	7.7	9.6	11.5
50	2.0	4.0	6.0	8.0	10.0	12.0
52	2.1	4.2	6.3	8.3	10.4	12.5

The true effects of inflation and taxes make the point that a simple statement of return without these negative impacts does not tell the story. The higher the combined federal and state tax rate, and the higher the rate of inflation, the more you need to earn (whether from investments or business activity) to maintain asset value and purchasing power. This raises another important and troubling issue: Does the inflation/tax risk mean you have to take greater risks to avoid losing purchasing power? Whenever your net return falls below the breakeven rate, you need to make some difficult decisions. These include:

1. Are the assumptions realistic concerning inflation? If you are applying an average CPI, does that include elements that do not directly impact your operation? If so, a more detailed study may reveal that a lower rate applies.

2. Does it make sense to pursue higher profits? If your current effective tax rate is minimally above the breakeven rate, what happens if higher profits translate to a higher rate? If the improvement in net profits results in a lower net return that is under the breakeven rate, does it even make sense to pursue those higher profits? Avoiding greater profits due to taxes is counter to the basic capitalistic model of commerce, but reality cannot be ignored.

3. Can the situation be improved by shifting domicile and markets? For example, if your company pays taxes in New York, New Jersey, or California, can you take

steps to reduce the state tax burden? Can you relocate to Texas, Washington, South Dakota or Nevada? The question is more complex than simply changing headquarters; the state where revenues are generated is normally where you pay taxes. Consequently, moving your offices to a no-tax state does not solve the problem completely. The issue of pursuing a different geographic customer is more complex and difficult but should be considered in the mix of possible solutions as well.

4. Is it possible to increase your rate of return, whether on short-term investments or on long-term business activity? In the weak economy of 2008 and 2009, many publicly listed companies were able to improve net profits even when revenues were lower than the previous year. This was accomplished by cutting expenses which, in most situations, meant cutting jobs. This is not a good idea if you also want to grow; however, applying austerity to the income statement may be aimed at achieving a higher net profit (net of taxes and inflation) without cutting off expansion. There are invariably methods available to reduce expenses, but only to a degree. The issue always comes down, eventually, to a decision about appropriate risk levels and margins of profit.

5. Can the margin of profit be improved? This is a difficult task and requires taking more risks in the marketplace. How can your competitive stance be improved without exposing your employees, stockholders, and vendors to higher market risk? Margin of profit can be improved through greater efficiencies in your supply chain, mergers with competitors, or taking greater market risks. These are the issues your company must confront when it finds the current level of net return to be lower than the breakeven return rate.

Calculating Cash Flow

Managers are preoccupied with cash flow, more than with net profits. There are good reasons for this. If your working capital is not strong enough to pay current bills on time or to finance expansion when opportunities arise, then profits showing up on paper are useless. Cash is much more important than profits.

Example: A company reports annual profits of $1.6 million on sales of $20 million in revenues, or 8%. However, current bills are chronically past due. This is caused by several factors: Inventory levels are too high, accounts receivable are being collected slowly, and the company has several loans outstanding.

When a *combination* of mitigating factors erodes working capital, it has negative consequences. Slow payment of current bills leads vendors to place companies on COD status or even to pursue collection, further eroding a company's credit. This makes it impossible to get further financing. Meanwhile, current obligations for repayment of principal and interest are demanding a growing percentage of profits each month. The tendency is that once cash flow begins to deteriorate, it gets worse.

To calculate the dollar amount of cash flow each month, begin with the reported net profit; and then add in non-cash expenses and other sources of funds; and deduct all payments not reflected in the income statement. Non-cash expenses include depreciation as the most important adjustment. If prepaid accounts and reserves have been set up, the current year's expense entries to those accounts should be added back in as well. Other sources of funds can include proceeds from:
- new loans approved and granted
- sales of capital assets
- legal settlements received

Non-income declines in cash flow include payments made for:
- purchase of capital assets
- repayment of loan principal
- dividends declared and paid
- legal settlements paid
- payments to reduce current liabilities

The formula for calculating net cash flow is:

Formula: Cash Flow

$$I + (N + L + A + S + O) - (L + A + S + D + O) = C$$

where: I = net income
N = non-cash expenses
L = loan transactions
A = capital asset transactions
S = legal settlements and judgments
O = other adjustments
D = dividends paid
C = cash flow

Example: Your company has reported current year net profits of $846,500. This includes depreciation, $42,000; loan proceeds, $10,000; the sale of capital assets, $16,000; receipt of $5,000 as a legal settlement; and $2,000 for other adjustments, increasing cash. It also includes loan principal repayments of $114,000; purchase of capital assets, $307,000; payment of legal judgments, $265,000; dividends paid, $110,000; and other adjustments, $15,000. Cash flow is calculated as:

$$\$846,500 + (\$42,000 + \$10,000 + \$16,000 + \$5,000 + \$2,000)$$
$$- (\$114,000 + \$307,000 + \$265,000 + \$110,000 + \$15,000) = \$110,500$$

Even though the year's profits were $846,500, the net cash flow picture grew by only $110,500. The adjustments are significant because in this situation, management might be considering an expansion plan. The argument might be put forth that because profits were $846,500, the estimated cost of expansion of $600,000 is easily afforded. When cash flow is reviewed with the adjustments, however, the flaw in this conclusion is apparent. Because there were so many negatives involved, the company's cash flow is not adequate to fund the expansion plan.

The calculation of cash flow can be done on a spreadsheet, with the following values:

A1 I
B1 N
B2 L
B3 A
B4 S
B5 O
B6 =SUM(B1:B5)
C1 L
C2 A
C3 S
C4 D
C5 O
C6 =SUM(C1:C5)
D6 =SUM(A1+B6-C6)

These field values, when entered based on the preceding example, produce the following results:

A	B	C	D
$846,500	42,000	$114,000	
10,000		307,000	
16,000		265,000	
	5,000	110,000	
	2,000	15,000	
	75,000	811,000	$110,500

Cash flow calculations affect your ability to expand operations or to reduce debt levels, not to mention keeping current obligations current. The calculations should be part of the budgeting process, with expenses expressed in the form of the coming year's budget, revenues in the forecast, and cash flow in the projection. These variations in terminology help to keep clear the distinctions between the groups of activity being estimated within the overall budgeting process.

The analysis of return and profit on a realistic basis—considering both inflation and taxes as part of the study—is not just an exercise. In preparation of revenue forecasts, expense budgets and cash flow projections, adjusting for inflation and tax assumptions aids in development of fact-based reports. It also helps to better understand current returns from operations or investments. If the operating profit or investment return is not understood in terms of how purchasing power of money is impacted, any analysis (even a comparative analysis) is lacking.

The next chapter expands on the concept of using ratios to study the balance sheet; evaluate working capital; and compare financial and capital strength. The analysis of the balance sheet as a test of your company's basic financial condition is a sensible way to monitor and score overall performance and the potential for expansion.

Chapter 5
Financial Reporting Formulas: The Balance Sheet

The two primary financial statements used in business are the balance sheet and the income statement. This chapter explains how the balance sheet is put together and provides formulas for the most useful balance sheet ratios.

Even managers who have not had accounting education can master the short list of calculations that reveal the status and financial condition of a business. You do not need to get involved with a highly technical level of analysis; however, you do need to know about one dozen formulas that can be applied and compared between companies or over time.

Balance Sheet Basics

The balance sheet gets its name from two of its most important features. First, this report contains the balances of all asset, liability and net worth accounts as of a specific date. This date is the end of a fiscal quarter or year. The date also corresponds to the ending date of the period for which the same period's income statement reports. For example, the income statement reports revenues, costs, expenses and profits for 12 months from January through December 31. The balance sheet released for the same period reflects balances on December 31.

The second source for the name is that the report summarizes a balance between the accounts, based on a formula:

Formula: Balance Sheet

$$A = L + N$$

where: A = assets
L = liabilities
N = net worth

Example: For example, your company's year-end report includes $14,607,500 in assets. The balance sheet also reports $11,477,250 in liabilities and $3,130,250 in net worth. The formula, applied to these values, is:

$$\$14,607,500 = \$11,477,250 + \$3,130,250$$

The formula can also be summarized on a spreadsheet program using the following values:

DOI 10.1515/9781547400638-005

A1	A
B1	L
C1	N
D1	=SUM(B1+C1) (should match cell A1)

By definition, "assets" includes all properties owned by the company: cash, accounts receivable, and capital assets, for example. Liabilities are the debts held by the company. Net worth consists of capital stock in corporations (or owners' equity in partnerships and sole proprietorships) as well as any accumulated earnings from past years (called retained earnings). Corporations reduce net worth when they pay dividends, and any repurchase of outstanding shares is also booked into the net worth account as Treasury Stock.

However, the balance sheet is not necessarily a fully accurate document; any mathematical reliance on reported numbers must be accepted. For example, some liabilities may be left off the balance sheet altogether. Off-balance sheet financing of debt may also involve off-balance sheet debt along with its interest expense, even when that interest expenses is deducted on the firm's tax return:

> The idea that although firms may have incentives to keep debt off their balance sheet, they also have incentives to include interest expense of their tax returns ... it is common to structure off-balance sheet financing arrangements so that the firm received the benefits of the interest deduction for tax purposes while avoiding reporting the obligation and the interest expense on its financial statements.[1]

To the non-accountant, this distortion will appear to artificially change reported liabilities as well as taxable income. It does; but under generally accepted accounting principles or GAAP rules, there is plenty of room to adjust like this to make the asset, liability, and net worth sections of the balance sheet appear as favorably as possible.

The paradox of accounting rules is only one of many problems with the balance sheet. How the balance sheet balances itself is a mystery to most people who have not been trained in double entry bookkeeping. The system, with one debit (left side) and one credit (right side) for every transaction, ensures that any math errors made in recording will show up because the journals will not balance (meaning debits and credits net out to zero).

This system is traced back to 1494, when Friar Luca Bartolomeo da Pacioli developed a system he called *alla Veneziana*, which is used universally today and has been renamed double-entry bookkeeping:

---

1 Maydew, E. (2005). Discussion of Firms' Off-Balance Sheet and Hybrid Debt Financing: Evidence from Their Book-Tax Reporting Differences. *Journal of Accounting Research, 43*(2), 283–290.

In their ledgers, Venetian merchants separated debits and credits, dividing them into two columns. As Pacioli wrote: "All the creditors must appear in the Ledger at the right-hand side, and all the debtors at the left. All entries made in the Ledger have to be double entries—that is, if you make one creditor, you must make someone debtor."[2]

Debits are added, and credits are subtracted. Some typical entries employing this system:

Description	Debit	Credit
Sales on account	Accounts receivable	Sales
Payment on account	Cash	Accounts receivable
Current debt obligation	Expense	Accounts Payable
Payment of the debt	Accounts payable	Cash

This process applies to every transaction made in the books of the company. At the end of the period, the total of all debits is equal to the total of all credits. When the net total of all balance sheet accounts (assets, liabilities, and net worth) is added up, the balance should be equal to the net total of all income statement accounts (revenue, costs, expenses). This equal balance on each side is the net profit. The income account balances are zeroed out when the books are closed, and the profit is credited (or the loss debited) into the net worth section of the balance sheet. This creates the perfect balance of the balance sheet.

Valuable Resource: For a detailed explanation of the double entry system, check http://simplestudies.com/accounting/lessons/p0401.htm

---

The components of the balance sheet are broken down into several sub-categories, which become very important when balance sheet ratios are developed. These sections are:
- *Current assets*, which are assets in the form of cash or that are convertible to cash within one year. The current assets include cash, accounts receivable (net of a reserve for bad debts), notes receivable, marketable securities (investment accounts), and inventory.
- *Capital assets*, also called long-term or fixed assets, are the buildings, vehicles, equipment and machinery owned by the company, less accumulated depreciation.
- *Intangible assets* are any assets without physical value, including goodwill and covenants not to compete, for example.
- *Prepaid and deferred assets* are accounts set up to be written off over several years. For example, a three-year insurance premium is set up as a "prepaid" asset

---

2 How a Medieval Friar Forever Changed Finance. (2012). *The Accounting Historians Journal, 39*(2), 113–115.

with one-third moved to the expense account each year. An expense paid before it is due is set up as a deferred asset, to be reversed and moved to the expense account in the proper accounting period.

- *Current liabilities* are obligations that are payable within 12 months. These include accounts payable, payroll and other current taxes, and 12 months' payments on long-term notes.
- *Long-term liabilities* are all debts not payable within the next 12 months. These include long-term notes or bonds.
- *Deferred credits* are timing differences like deferred assets. For example, if a company receives payment for revenue that will not be earned until the following year, it is properly set up as a deferred credit and later reversed and moved to the revenue account.
- *Capital stock* is the initial value of stock issued and outstanding. This value does not change unless additional shares are issued, or until the company buys its stock and retires it permanently as Treasury Stock.
- *Retained earnings* is an account that accumulates each year's profits (as an addition) or losses (as a subtraction) in the net worth section.
- Other net worth accounts include non-deductible expenses and dividends.

The purpose of deferral and prepaid accounts is to ensure that all transactions are "booked" into the right earnings period. One purpose to the accrual system of accounting is to manage transactions to achieve this goal. For example, sales on account (which for many companies are most of sales) should be booked into the month when the sale is made, even if payment will not be received for several months. A liability is booked as an accrual in the period when the goods are received, even if you will not pay for them for 30 days. The process of accruals and reversals is the most complex aspect of bookkeeping, even though it is a necessity. Today, with mostly automated entry systems in use, most managers do not need to be concerned with these complexities, except in one respect: The accrual process is not only the area which is the most complex, but also the process where manipulation most often occurs. If a company wants to "cook the books" by over-reporting revenue or deferring expenses, the accrual system can be used to distort the truth. One important service that ratio analysis provides is in uncovering any suspicious or questionable trends. The trend reveals everything, including distortion of what is taking place. For this reason alone, every manager can vastly improve the ability to study financial reports by applying a few ratios and tracking the trends they represent.

The practice of cooking the books is disturbingly common. Even with independent audits conducted to ensure accuracy in the balance sheet and other financial reports, it is remarkably easy for companies to intentionally inflate earnings through practices such as "channel stuffing." This is:

... a deceptive business practice of inflating sales by deliberately sending to the retailers along its distribution channel more products than they are able to sell in the normal course of business and allowing the retailers to pay later ... the accounting system records the increase in receivables as an investment. Because the income statement is based on all sales—whether collected during the period or not—channel stuffing leads to higher current income at the expense of lower future income.[3]

The many ways that companies can change their balance sheets to create improved earnings and higher equity, point out that being able to calculate balance sheet ratios is only half of the challenge. The greater question is whether the math used at the corporate level is based on reliable numbers.

Working Capital Ratios

Balance sheet account balances are used to track some important ratios, especially concerned with working capital (the availability of funds) and capitalization trends (the balance between equity and debt as capital sources within the organization). The first of these two major areas, working capital, involves a few important ratio tests. First and best-known is the current ratio.

This is a comparison between total current assets and total current liabilities. Both groupings of accounts relate to the availability and use of cash and cash equivalents within the next 12 months. The current ratio is supposed to serve as a test of how well a company can pay its debts over the next 12 months. The current ratio is calculated by dividing current assets by current liabilities:

Formula: Current Ratio

$$A \div L = R$$

> where: A = current assets
> L = current liabilities
> R = current ratio

Example: A company reports that its current assets at the end of the year are $415,300 and that its liabilities are $207,900. The current ratio is:

$$\$415,300 \div \$207,900 = 2.0$$

---

3 Livnat, J., & Santicchia, M. (2006). Cash Flows, Accruals, and Future Returns. *Financial Analysts Journal, 62*(4), 48–61.

The formula can also be reduced to a basic spreadsheet program, with the values entered as follows:

A1 A
B1 L
C1 =SUM(A1/B1)

As a general observation, a current ratio of 2.0 or higher is a good sign, indicating that liquidity is in good shape and that the company can fund its debts. A ratio below 1.0 (meaning current assets are lower than current liabilities) is a warning sign, an indication that the level of working capital is a problem and that the company might have problems meeting its obligations in a timely manner. As with all ratios, the current ratio needs to be tracked over a period of quarters or years to see where the trend is headed.

Some problems are possible in trying to accurately interpret the current ratio without analyzing its components. For example, if the company is carrying too much inventory, or if it is accumulating accounts receivable without timely collections, the current ratio can appear healthy when, in fact, there are emerging problems in managing working capital. The current ratio does not tell the entire story.

The current ratio can also be manipulated to achieve a desired level. In fact, "the current ratio normally has little value ... since its computation usually results in a meaningless numerical figure. If not used with caution it can even be misleading in the majority of cases."[4]

This makes the current ratio further questionable as an indicator. For example, a company has the following values just before closing its books:

Current assets	$702,300
Current Liabilities	$412,300

Given these levels, the current ratio will be 1.7 ($702,300 ÷ $412,300). However, if the company pays off $75,000 in current liabilities just before closing the books, the current asset and liability levels are changed to:

Current assets	$702,300 – $75,000 = $627,300
Current liabilities	$412,300 – $75,000 = $337,300

Now the current ratio is 1.9, close to the desired 2.0 you look for in the current ratio. Once the current balance sheet has been released, the company can simply allow current debts to rise once again to replace the $75,000 spent. In this situation, the truer picture was distorted by timing payments for some obligations, also hiding the real cash flow picture from the analyst.

---

4 Bowlin, O. (1963). The Current Ratio in Current Position Analysis. *Financial Analysts Journal*, 19(2), 67–7.

Because the current ratio is useful, especially as a long-term tracking device for working capital, it should be maintained in your analytical arsenal, even with its shortcomings. However, to best deal with the possible problems not revealed by the current ratio, you may also want to track the quick assets ratio, also called the "acid test." To calculate:

Formula: Quick Assets Ratio

$$(A - I) \div L = R$$

> where: A = current assets
> I = inventory
> L = current liabilities
> R = current ratio

Example: The company you are studying reported $415,300 in total current assets, including $206,600 in inventory; and $207,900 in liabilities. The quick assets ratio:

$$(\$415,300 - \$206,600) \div \$207,900 = 1.0$$

The formula can be summarized on a spreadsheet with the following cell values entered:

```
A1  A
B1  I
C1  L
C1  =SUM(A1-B1)/C1
```

With the exclusion of inventory, the quick assets ratio is expected (as a general standard) to reside at or above the level of 1.0. This adjustment is valuable in situations where the inventory level is going to vary greatly from one season to another. A retail operation, for example, would be expected to have a very high inventory level at the end of the third quarter in anticipation of the holiday season; and a very low inventory level at the end of the first quarter. Because these variations are caused by cyclical factors and not as indications of changes in the business climate, the quick assets ratio can serve as a more reliable quarter-to-quarter predictor than the better-known current ratio.

This ratio is also more reliable because inventory is not as easily converted to cash as other forms of current assets. Even when inventory is depleted through sales activity, it may be first converted to accounts receivable before cash is finally generated. These realities make the quick assets ratio a worthy accompaniment to the current ratio.

Ratios for Management of Working Capital

The current ratio and quick assets ratio provide general indicators of how well working capital is generated. Such ratios should be tracked over many quarters to see how a trend is developing. Going beyond these initial tests, you can also decide how effectively management is using its cash, with two specific ratio tests. The previous section explained why testing overall current assets may not reveal all you need to know. Inventory or accounts receivable levels can be allowed to climb too high. Testing these is an important attribute of a working capital analysis. The following discussion includes an explanation of how to test accounts receivable; a useful inventory ratio test appears later in this chapter.

To test accounts receivable as one component of working capital, you need to decide how effectively the company is collecting its outstanding balances. If the balances are being allowed to remain outstanding for too long a period, then the trend is negative. One very basic way to track this is to "age" the current accounts receivable balances. To age the account, each company owing money is listed on a worksheet and then shown based on how many days the balance has been outstanding. Those between zero and 30 days are current; those between 31 and 60 days are later but not yet past due. Any accounts over 61 days are past due, and those over 91 days are seriously past due.

Account managers know that the longer accounts remain unpaid, the less likely it becomes that you will ever receive money. Even those accounts that are eventually collected adversely affect cash flow because of the length of time required to make collections. The aging list can be set up based on the formal shown in Table 5.1.

The aging list helps define the time required to make collections, especially when the list is divided into percentages. When the time required is growing, a company can act to curtail the problem by:

1. Immediately cutting off credit to any company more than 60 days past due.
2. Improving the initial screening process for customers requesting credit, including a review of credit reports for new customers.
3. Accelerated contact policies starting with direct telephone calls to ask for payments.
4. Imposing credit limitations for high-volume customers, especially those that are slow to make payments.

Table 5.1: Accounts receivable aging list

Name	Total	31–60	61–90	over 90
Percentage	**100%**	%	%	%

These steps will prevent the problem from worsening and may even prevent future bad debts. However, the overall credit and collections situation can be distorted by only a few high-volume customers. In periods when sales volume is rising quickly, the relationship between granting of credit and collections tends to change as well. One way to keep control over the problem of working capital during an expanding market period is to track the days' sales outstanding. This is a ratio that tracks sales on a full-year moving basis (continually updating the full 12 months' sales total) and then comparing that average to the changing balances of accounts receivable. To calculate, first determine the revenue total for the past full year. Restrict this to those sales made on account, excluding cash paid at the time of the sale. This requires maintaining a running tally. Each day's new credit-based sales are calculated by adding the latest day and dropping the oldest day. To make this calculation easier, you can also track sales on a 52-week basis. In this alternative, add the latest week's sales to the running total and subtract the oldest week. This can also be applied on a 12-month level, adding last month's sales to the running balance and then subtracting the oldest one, so that you continually end up with a full year's updated revenue total. Then apply this information in the formula:

Formula: Days' Sales Outstanding

$$R \div (S \div 365) = D$$

where: R = Accounts receivable balance
S = one year's sales on credit
D = days' sales outstanding

Example: Your company's sales on credit for the past full year have been $9,415,800. The current balance of Accounts receivable is $1,042,700. To calculate the days' sales outstanding using the formula:

$$\$1,042,700 \div (\$9,415,800 \div 365) = 40$$

This reveals that the average Account remains outstanding for 40 days. Given that some are paid much faster and some remain outstanding for much longer, this outcome is not surprising. As a matter of judging how effectively outstanding bills are being managed, you would hope to see this number fall as steps are taken to speed up collections. When the number begins to rise, notably during periods of revenue expansion, it is a warning sign that working capital is not being managed well.

The formula can also be reduced to a spreadsheet program using the following values:
A1   R
B1   S
C1   =SUM(A1)/(B1/365)

Another valuable test of working capital management is interest coverage. This tests the cash available before paying interest and taxes, versus the interest a company must pay on its existing debt. This "margin of safety" in cash levels is at times used by lenders to determine how well a company can repay a loan. The calculation begins with EBITDA, an acronym meaning "earnings before interest, taxes, depreciation and amortization." Put another way, EBITDA is the cash-basis operating income. It excludes non-cash expenses (depreciation and amortization) and is the value you find before deducting interest expense and taxes. The formula for EBITDA is:

Formula: EBITDA

$$N - (I + T + D + A) = E$$

where: N = net income
I = interest expense
T = taxes
D = depreciation
A = amortization
E = EBITDA

Example: A company reports the bottom line net income of $1,680,000. However, this includes interest expense of $115,200; taxes of $712,300; depreciation of $186,000; and amortization of $2,400. EBITDA is:

$1,680,000 - ($115,200 + $712,300 + $186,000 + $2,400) = $664,100$

On a spreadsheet, calculate EBITDA by entering:
A1 N
A2 I
A3 T
A4 D
A5 A
A6 =SUM(A2:A5)
B1 =SUM(A1-A6)

The calculated EBITDA is a problematical value in financial analysis. It is going to vary considerably based on the level of depreciation and taxes, for example, making comparisons between different companies very difficult. It has been described as "a fairy tale told to investors and credit managers so that they go to sleep happy instead of running for the hills."[5]

However, for calculating interest coverage as a test of operating cash-based income versus interest expense, it is a useful component in the formula:

Formula: Interest Coverage

$$E \div I = C$$

> where: E = EBITDA
> I = interest expense
> C = interest coverage

Example: Using the previously calculated values, interest coverage is:

$$\$2,695,900 \div \$115,200 = 23.4 \; times$$

This can also be summarized on a spreadsheet program:

A1　E
B1　I
C1　=SUM(A1/B1)

The ratio is expressed as the number of times interest expense represents EBITDA. This is a valuable calculation when a company's ability to repay is compared to a standard level used by a lender, as well as to the changes this ratio produces as the working capital picture changes over time.

Capitalization Ratios

One of the most overlooked areas of financial analysis is capitalization. This is a reference to the types of financing a company uses to fund operations and growth. There are two types of capitalization: equity and debt. Equity capitalization is the value of issued and outstanding stock and retained earnings, and debt capitalization consists of long-term loans, notes and bonds—in other words, borrowed money.

---

5 Gavin, Ted. (December 11, 2011). Top Five Reasons Why EBITDA Is A Great Big Lie. *Forbes,* at https://www.forbes.com/sites/tedgavin/2011/12/28/top-five-reasons-why-ebitda-is-a-great-big-lie/#2f48dfa6d2da, retrieved June 25, 2018.

Important differences are found between short-term (current) debt and long-term debt, not only in how these debts are repaid, but in how they are treated in financial analysis. As another example of potential manipulation of the books, some companies have reclassified short-term debt as long-term, and then reverted it back to original status. When this is done, the company artificially controls how it reports its capitalization and leverage levels: "Although aggregate measures of liabilities and equity remain unchanged when firms reclassify (declassify), the practice does increase (decrease) reported measures of liquidity, such as the current ratio, and long-term leverage."[6]

Such manipulations distort the reported status of balance sheet items. The higher the level of debt capitalization, the higher the level of profits that must be spent in debt service, the repayment of loans. The higher the interest a company must pay, the less capital remains for expansion and dividends. As debt rises, the relative impact of equity declines, and stockholders have less chance of future growth. From a manager's point of view, increasing debt means more negative impact on cash flow. At some point, the level of debt can take such a prominent role that it becomes impossible to ever reverse the course.

Growing levels of debt are a huge problem for any company interested in growth and in maintaining a return on revenues and equity. For this reason, analyzing and tracking the trend in total capitalization is one of the most important and revealing tests of a company's health that you can perform. In the role of manager, controlling debt is simply good cash flow control. As an investor or potential investor, observing the long-term trend tells you whether the company is growing in positive directions or allowing itself to be gradually consumed by growing debt.

The most effective test of how a company is financing its operations, and of the trend underway, is by monitoring the debt to capitalization ratio. This is a test of the level of debt as a percentage of total capitalization. The formula tracks this on a percentage basis:

Formula: Debt to Capitalization Ratio

$$D \div T = R$$

where: D = long-term debt
T = total capitalization
R = debt to capitalization ratio

6 Gramlich, J., McAnally, M., & Thomas, J. (2001). Balance Sheet Management: The Case of Short-Term Obligations Reclassified as Long-Term Debt. *Journal of Accounting Research, 39*(2), 283–295.

Example: A company you are tracking reports total capitalization (in millions of dollars) of $42,950. Long-term debt is $13,030. The debt to capitalization ratio is:

$$\$13,030 \div \$42,950 = 30.3\%$$

This ratio can also be tracked on a spreadsheet using the following cell values:

A1   D
B1   T
C1   =SUM(A1/B1)

The example demonstrates that the company is employing one-third long-term debt and two-thirds equity to finance operations. Is 33% too high? That depends on the industry standard and other factors. For example, if the company is in an industry demanding maintenance of high inventory levels or purchase of capital assets every year, a relatively high debt to capitalization ratio must be expected. The most important test is the trend, and not only one year's debt to capitalization ratio. Is the company maintaining the same level (or a declining level) of debt as a percentage of total capitalization each year? Is the relationship between debt and equity remaining the same or declining while revenue and profits are growing? One danger sign is a growing debt to capitalization ratio during periods of expansion. The worst case is one in which revenues are growing every year and the debt to capitalization ratio is also growing, while net profits are remaining the same or declining. That is a sign of poor cash flow management.

The debt to capitalization ratio also should be tracked along with the current ratio as a comprehensive test of management and its cash flow policies. Knowing that investors and analysts pay a lot of attention to the current ratio while too often overlooking the debt to capitalization ratio, it is tempting to use long-term debt to artificially bolster the current ratio. This creates the impression that the company is managing cash flow effectively when it is losing control.

Example: A corporation has reported net losses in the past three years but has managed to maintain a current ratio of 2. Based on this, some analysts are confident that management is maintaining effective working capital levels. But a more in-depth analysis reveals the opposite. The numbers (in millions of dollars) reveal:

Year	Net loss	Current assets	Current liabilities	Long-term debt	Total capitalization
2016	(4,889)	16,500	8,007	33,006	101,550
2017	(2,006)	27,455	13,550	44,615	105,153
2018	(10,588)	41,060	20,625	56,673	101,623

An initial analysis, limited only to a study of current ratio trends, reveals the following positive outcome:

Year	Current assets	Current liabilities	Current ratio
2016	16,500	8,007	2.1
2017	27,455	13,550	2.0
2018	41,060	20,625	2.0

This outcome meets the criteria for the current ratio and at this point, many analysts would end their analysis of cash flow control. However, when you study the debt to capitalization ratio for the same period, you find:

Year	Long-term debt	Total capitalization	Ratio
2016	33,006	101,550	32.5
2017	44,615	105,153	42.4
2018	6,673	101,623	44.8

This analysis demonstrates that the debt to capitalization ratio moved significantly higher even while the current ratio remained steady. During periods of net losses, you would not expect to be able to maintain as healthy an outcome, but something else was going on in this example. The level of current assets rose at about the same dollar amount as long-term debt each year. This means the company was committing itself to long-term loans or bonds and using funds to increase current assets or decrease current liabilities, maintaining the current ratio level even while long-term debt was growing.

Without a specific example of how long-term debt and current ratio can be controlled, this explanation might not seem relevant. However, examples can be found in historical financial reports of listed companies. Looking back to the period between 2013 and 2018, Lockheed Martin (LMT) was able to hold its current ratio between 1.1 and 1.4. A study of the actual reported numbers shows how that this consistently occurred at the cost of higher long-term debt:

Year	Current ratio	Ratio
2013	1.2	55.6
2014	1.1	64.4
2015	1.1	77.9
2016	1.2	89.9
2017	1.4	99.0

By the end of 2017, nearly all capitalization for LMT was long-term debt. Only 1 percent of capitalization was in the form of stockholders' equity. However, anyone who checked only the current ratio would have been reassured that all was well.

This steady current ratio level was accomplished by increasing the long-term debt level; the debt to capitalization ratio grew from 55.6% to 99.0% during this period. This does not necessarily mean that Lockheed's management intentionally manipulated the outcome to maintain the ratio; and under the accounting rules, there is nothing illegal about increasing debt levels. However, the outcome kept the current ratio at an acceptable level, which was artificial when the long-term consequences are considered as well.

The conclusion of this exercise is that ratios should be studied over several years and in some cases (such as working capital and capitalization analysis) reviewed in conjunction with other financial ratios. The collective value of studying the debt to capitalization ratio *and* current ratio together is much greater than tracking either one in isolation.

A related test is intended to track a company's ability to make debt payments from current levels of net income. The debt coverage ratio is a test of both cash flow and capitalization. The formula compares the total of debt payments (including principal and interest) to net operating income for the same period. The formula:

Formula: Debt Coverage Ratio

$$I \div D = R$$

where: I = net income
D = debt service
R = debt coverage ratio

Example: Net income for a full year was recently reported as $12,415 and over the same period, the total of principal and interest on outstanding loans was $9,516. The debt coverage ratio was:

$$\$12,415 \div \$9,516 = 1.31$$

Lenders use the debt coverage ratio and its trend over time to judge how well a company can afford repayments of loan commitments. This ratio, like all others, should be tracked over many years to judge whether the situation is getting better or worse. The lower the factor, the worse the cash flow health of the company (for example, if 100% of net profits were used to make loan payments, the debt coverage ratio would be 1.00).

The formula can be tracked using a simple spreadsheet program:
```
A1   I
B1   D
C1   =SUM(A1/B1)
```

A final ratio is called the liability-to-asset ratio. This involves dividing total liabilities by total assets. This provides an interesting variation on both current ratio and debt to capitalization ratio; but it is not widely used because it involves *all* assets and liabilities; any change in an account will affect and perhaps distort the trend. For example, if the company sells a large capital asset for more than book value, the total value of assets is increased (cash is received above the equipment's book value). At the same time, any remaining liability for the equipment is instantly eliminated. If the relative levels of assets and liabilities remain the same, the formula can be a useful indicator, but not as valuable as current ratio combined with debt to capitalization ratio. The outcome of this ratio shows how well the level of assets are funded by liabilities (the remaining level would be funded by equity). In this regard, the liability-to-asset calculation is like the debt to capitalization ratio. The formula:

Formula: Liability-to-Asset Ratio

$$L \div A = R$$

> where: L = total liabilities
> A = total assets
> R = liability-to-asset ratio

Example: A company's total assets are reported as $1,443,016 and its liabilities are $982,663. The liability-to-asset ratio:

$$\$982,663 \div \$1,443,016 = 68.1\%$$

The percentage of debt-based funding is 68.1%; this is not the same as capitalization, however, which compares only long-term liabilities and equity capital.

The formula can also be summarized on a spreadsheet:
```
A1   L
B1   A
C1   =SUM(A1/B1)
```

Combined Ratios

In addition to ratios using only balance sheet accounts, many financial ratios combine balance sheet accounts in comparison with the results reported on the income statement. Two turnover ratios are noteworthy. These are inventory turnover and a comparison between sales and capital (fixed) assets.

Inventory turnover compares the *average* inventory for the year to the cost of goods sold for the entire period. In developing average inventory, the level of change during the year should dictate how it is calculated. For example, if the inventory levels remain consistent throughout the entire year, a beginning and ending balance can be added together and divided by two. When levels change significantly depending on the calendar cycle, use either quarterly inventory levels or even monthly levels, depending on the level of change.

The ratio estimates the level of cash committed to maintaining required inventory levels, with the intention of indicating when levels should be reduced to improve cash flow. If the company is maintaining inventory at a higher level than required, it not only ties up cash; it also is spending too much money on transportation, insurance, property taxes, warehouse labor, recordkeeping, interest, and the danger of deterioration, damage, theft, and obsolescence. Keeping inventory turnover at a *minimum* required level based on sales volume for the season saves both cash flow and profits. The formula:

Formula: Inventory Turnover

$$C \div ((B + E) \div p) = T$$

where: C = cost of goods sold
B = beginning inventory
E = ending inventory
p = number of periods in the average
T = inventory turnover

Example: A company has reported cost of goods sold for the past fiscal year of $4,500,600. Inventory level at the beginning of the year was $702,400 and at the end of the year at $785,700. Determining that the monthly changes in inventory were not seasonally great, the calculation is performed using beginning and ending inventory only:

$$\$4,500,600 \div ((\$702,400 + \$785,700) \div 2) = 6 \text{ times}$$

This calculation concludes that the average inventory was "turned" (replaced) six times during the year. This also means that on average, the company maintains two months' worth of inventory on hand. This calculation is valuable as a means of tracking inventory management. If the number of turns begins to decline, it means inventory levels are rising relative to the cost of goods sold. If levels of two months' worth of goods are considered reasonable, a turnover rate below 6 times would be a negative trend.

The formula can be summarized in a spreadsheet program using the following cell values:

A1　C
B1　B
B2　E
C2　=SUM(A1/((B1+B2)/p))

In the preceding example, the cell values are:

A1　4,500,600
B1　702,400
B2　785,700
C2　=SUM(A1/((B1+B2)/2))

If you determined that inventory needed to be averaged in greater detail, the 'B' column would include quarterly inventory levels in cells B1 through B4, and the value of 'p' would be 4. If monthly averages were to be used, the monthly inventory totals would be entered in cells B1 through B12, and 'p' would be 12.

A second popular combined ratio is fixed asset turnover. This ratio analyzes a relationship between capital investment and sales generation. It is only applicable in companies in which a direct relationship exists between capital investment and generation of income. An accounting firm would not use this ratio to track its invest-ment in office equipment and furniture. However, a contractor would be likely to use it to track developments in its investment in equipment that is used on the job. The ratio is applied to judge whether investment over time is profitable. For example, if the turnover of sales to fixed assets begins to decline, it indicates a trend toward less efficiency. The calculation:

Formula: Fixed Asset Turnover

$$S \div ((B + E) \div p) = T$$

where: S = sales
B = beginning fixed asset value
E = ending fixed asset value
p = number of periods in the average
T = fixed asset turnover

Example: Your company reported sales last year of $8,005,400. The account value of fixed assets at the beginning of the year was $1,008,900 and at the end of the year, $1,188,000. Using only two periods per year to find the average value of fixed assets, the formula for fixed asset turnover is:

$$\$8,005,400 \div ((\$1,008,900 + \$1,188,000) \div 2) = 7.3$$

Fixed assets were turned an average of 7.3 times during the year, meaning that the sales level generated by capital investment was 7.3 times that investment. If the number of turns were to begin to decline, it would indicate reduced efficiency resulting from over-investment in fixed assets. However, this conclusion is relevant *only* if the nature of the company and type of sales can identify a direct relationship between fixed assets and sales generation.

The formula can be reflected using a spreadsheet program, with cell contents very similar to those for inventory turnover. In the case of fixed asset turnover, cell contents are:

A1	S
A2	B
A3	E
A4	ending inventory
A5	number of periods in the average
A6C2	=SUM(A1/((A2+A3)/p))

If you prefer using a greater number of asset values for the year, the inventory cells can be adjusted as well as the value of 'p.' For example, if you purchased or sold substantial portions of overall fixed assets during the year, it would be necessary to create an accurate average by using a greater number of account value entries.

The balance sheet ratios collectively summarize working capital status and trends, as well as overall capitalization. These are crucial trends to track. Combined ratios can also provide early warning signals when either inventory or fixed asset investment levels begin moving in a negative direction.

Continuing the analysis of ratios, the next chapter focuses on the income statement and explains a range of tests you can apply to evaluate profitability and internal controls.

Chapter 6
Financial Reporting Formulas: The Income Statement

While the balance sheet reports balances of asset, liability and net worth accounts at the end of the period, the income statement summarizes the transaction activities *during* the period. These include revenue, cost of goods sold, expenses, non-operating transactions, and net profit. The balance sheet and income statement are always released together; and the period covered by the income statement must end on the same date as the reporting date of the balance sheet. This is the end of a fiscal quarter or fiscal year.

The income statement (also called "profit and loss statement") includes accrued revenue, costs and expenses in most instances. Very few companies report on the cash basis because that does not reflect an accurate picture or status of the business. Accruals are essential to record all transaction in the proper period. Revenue must be shown during the period it was earned, and not in the period received; and costs and expenses must be shown in the period incurred, not in the period paid.

Example: A company reports major fourth quarter earnings over $16 million, but at the end of the year there remain outstanding about $11 million in accounts receivable. Reporting on a cash basis would under-report revenue by $11 million, because those sales were earned in the fourth quarter. Payment will not be received until later, but under the accrual system they are properly reported in the fourth quarter.

Example: At the end of the fiscal year, a manager notices that the accounts payable total is quite high, at $4,822,000. If the year's income statement were prepared on the cash basis, the expenses for the year would be under-reported by that amount. The expenses were incurred in the fourth quarter even though the bills will not be paid until later. Under the accrual system, the accounts payable are set up to reflect incurred expenses.

The accrual entries in these examples set up an accounts receivable offset by reported revenue, and an accounts payable liability offset by expenses:

Account	Debit	Credit
Accounts receivable	$11,000,000	
Revenue		$11,000,000
Expense accounts	$4,822,000	
Accounts payable		$4,822,000

These simplified account entries increase both current assets and current liabilities as offsets to the income statement entries for revenue and expenses. By booking (in

DOI 10.1515/9781547400638-006

accounting terms, "recognizing") these transactions with accruals in their proper periods, the income statement will be accurate.

The accrual system is essential to proper presentation of the range of transactions in the period covered by the income statement. However, because it is also possible to affect results by simply making an accrual journal entry, the methods of reporting must be carefully controlled and audited. Publicly traded companies are required to undergo audits and to certify that the final versions of financial statements are accurate. Without the independent oversight from an auditing firm, it would be possible to modify financial reports to look positive even when the true outcome is bleak. A premise for all managers in reviewing financial statements is to realize that accurate reporting relies on the integrity and thoroughness of the independent audit. This is not an absolute guarantee as history has proved (recall the case of Enron as well as dozens of other cases where audits failed to ensure accurate financial statements). However, even with its flaws the system in place for audit review is the best available method for preparation of financial reports. Even with the flaws in the U.S. system, it is probably the most reliable and accurate accounting process in the world.

Even so, the opportunities to "interpret" financial results make accrual accounting a problem regarding accuracy. This is especially true in a non-publicly listed company where the owner or manager can make accounting decisions without oversight from auditors, and is engaged in a principal-agent business model:

> A trade-off between the firm's financial reporting system and the contracting system determines whether the owner will find it optimal to induce the manager to report the firm's economic earnings or to manage the reported earnings by exercising reporting discretion. We refer to the former as "truthful reporting" and the latter as "earnings management," where earnings management incorporates any accrual-based manipulation of economic earnings by the manager.[1]

Income Statement Basics

An income statement is a series of summarized account totals for a specified quarter or year. The report moves from top to bottom, beginning with revenue and subtracting a series of costs and expenses. The sequence of reporting is:

Revenue
Less: Direct Costs
Equals: Gross Profit
Less: Expenses
Equals: Net Operating Profit

---

1 Evans, J., & Sridhar, S. (1996). Multiple Control Systems, Accrual Accounting, and Earnings Management. *Journal of Accounting Research, 34*(1), 45–65.

Plus or Minus: Non-Operating Income or Expense
Equals: Pre-Tax Net Profit
Less: Income Taxes
Equals: Net After-Tax Profit

This series of steps is used by virtually all companies. For example, the fiscal year 2017 report for Microsoft showed the following income statement results (in millions of dollars):[2]

Revenue	$89,950
Less: Direct Costs	55,689
Equals: Gross Profit	$34,261
Less: Expenses	11,935
Equals: Net Operating Profit	$22,326
Plus or Minus: Non-Operating Income or Expense	431
Equals: Pre-Tax Net Profit	$22,757
Less: Income Taxes	1,553
Equals: Net After-Tax Profit	$21,204

The components of the income statement can be further brown down into a series of formulas:

Formula: Gross Profit

$$R - C = G$$

where: R = revenue
C = direct costs
G = gross profit

In this chapter, the spreadsheet entries are provided using a vertical format rather than horizontal. This conforms to the income statement reporting format, from top to bottom. The spreadsheet entries for gross profit:

A1 R
A2 C
A3 =SUM(A1-A2)

---

2 Source: Microsoft 2017 Annual Report at www.microsoft.com

Formula: Operating Profit

$$- E = O$$

where: G = gross profit
E = expenses
O = operating profit

The spreadsheet entries:
A3 G
A4 E
A5 =SUM(A3-A4)

Formula: Pre-tax Net Profit

$$O + (-) N = P$$

where: O = operating profit
N = non-operating income or expense
(net income added, net expense deducted)
P = pre-tax profit

The spreadsheet entries in this situation include *subtracting* the net non-operating expense. In the case of a net income, it would be added:
A5   O
A6   N
A7   =SUM(A5-A6)

Formula: Net After-tax Profit

$$P - T = Z$$

where: P = pre-tax profit
T = income tax liability
Z = net after-tax profit

The spreadsheet components:
A7 P
A8 T
A9 =SUM(A7-A8)

The formula for the complete income statement can also be summarized as:

Formula: Income Statement

$$R - C = G$$
$$G - E = O$$
$$O - N = P$$
$$P - T = Z$$

where: R = revenue
C = direct costs
G = gross profit
E = expenses
O = operating profit
N = non-operating income or expense
P = pre-tax profit
T = income tax liability
Z = net after-tax profit

On the spreadsheet:

A1 R
A2 C
A3 =SUM(A1-A2)
A4 E
A5 =SUM(A3-A4)
A6 N
A7 =SUM(A5-A6)
A8 T
A9 =SUM(A7-A8)

Entries with the Microsoft example are:

A1 89,950
A2 55,689
A3 =SUM(A1-A2)
A4 11,935
A5 =SUM(A3-A4)
A6 431
A7 =SUM(A5+A6) (this is added because the net is positive)
A8 1,553
A9 =SUM(A7-A8)

Revenue	$89,950
Less: Direct Costs	55,689
Equals: Gross Profit	$34,261
Less: Expenses	11,935
Equals: Net Operating Profit	$22,326
Plus or Minus: Non-Operating Income or Expense	431
Equals: Pre-Tax Net Profit	$22,757
Less: Income Taxes	1,553
Equals: Net After-Tax Profit	$21,204

This format can be shifted over to the second column (B) and the first column used to label entries by name. An example of this is shown in Table 6.1, the result of changing calculation labels to 'B' in place of 'A.' The format can then be expanded to make comparisons between quarter and year, or between two fiscal years. It can also be used to calculate dollar and percentage outcomes, explained in the next section.

Table 6.1: Income statement

Microsoft Corporation
Income Statement
For the fiscal year ended June 30, 2017

	($ millions)
Revenue	$89,950
Less: Direct Costs	55,689
Gross Profit	$34,261
Expenses	11,935
Net Operating Profit	$22,326
Non-Operating Income or Expense	431
Pre-Tax Net Profit	$22,757
Income Taxes	1,553
Net After-Tax Profit	$21,204

Dollar and Percentage Reporting

Income statements are summarized in dollar values, and in the case of large dollar amounts, the dollar values are expressed in millions of dollars. This makes the information more digestible. Referring to the Microsoft example, the revenue of $89,950 (in millions) is easier to comprehend than the full dollar value (rounded) of $89,950,000,000.

Statements are further reduced to a shorthand reporting format using percentages. This enables you to review not only the dollar values but the relationship between the various components of the statement and the top line, revenue. The formula for reducing each component to a percentage form is:

Formula: Percentage of Revenue

$$C \div R = P$$

where: C = income statement component
R = revenue
P = percentage

On a spreadsheet (based on assumption that column B contains dollar values:
A1   income statement component
A2   revenue
A3   =SUM(A1/A2)

Applying this formula to the previous example, the outcome will look like the income statement in Table 6.2.

Table 6.2: Income statement with percentages

Microsoft Corporation
Income Statement
For the fiscal year ended June 30, 2017
($ millions)

Revenue	$89,950	100.0%
Less: Direct Costs	55,689	61.9
Gross Profit	$34,261	38.1
Expenses	11,935	13.3
Net Operating Profit	$22,326	24.8%
Non-Operating		
Income or Expense	431	0.5
Pre-Tax Net Profit	$22,757	25.3%
Income Taxes	1,553	1.7
Net After-Tax Profit	$21,204	23.6%

This combination of dollar values and percentages allows managers to use the two shorthand methods (abbreviated dollar values in millions, and percentages compared to revenue) to very quickly make judgments about the company's operating results. This judgment can be applied between:
-   Fiscal quarters
-   The quarterly and annual results
-   Two or more fiscal results

- Two or more companies in the same industry
- Two or more companies in different industries
- Actual results and the forecast of revenue or budget of expenses and net income

This multi-tiered analysis makes the process easier and faster. It also provides a means to quickly spot the red flags of an outcome that makes no sense. For example, if revenue has been growing each year by five percent and the current year shows a jump of 35 percent, what has taken place? This could be caused by manipulation of the results (such as booking revenue too early, for example). However, the explanation could also be more benign. If the company acquired another organization during the year, all results (revenue, costs, expenses and profits) are going to depart from an established trend.

The unusual outcome can also be caused by an accounting valuation change during the year, or by a one-time accounting adjustment, called an "extraordinary" item. Whenever you see unusual or unexpected results to any part of the income statement, whether positive or negative, an explanation should be sought. If the explanation makes no sense or is not offered, that is the worst type of red flag, indicating deception. This is a rare event, but it can occur. You are more likely to be satisfied with an explanation.

An examination of trends is a valuable management tool, and it requires review of several years' results. You need to look at three years minimum, although the longer the period you review, the better. If 10 years of income statements are available, that gives you a healthy period to study the trends in revenue, costs and expenses and profits. Some cautionary points about trends:

1. *Trends are invalidated by mergers or sales.* Any trend involving revenue and profits remains valid only if the company's status remains unchanged. This means the same subsidiaries are owned throughout the period and there are no mergers or sales. As soon as an operating segment is sold, all existing trends must be recalculated excluding that segment. Any mergers or acquisitions also require recalculation of trends to ensure that the outcome is valid.

2. *Trends tend to level off over time, so maintaining the same level of growth is not likely.* Statistically, no trend is going to continue in a straight line indefinitely. A year-to-year increase in dollar amounts of revenue or net profits might continue; but over time that will represent a diminishing percentage increase over the previous year. This occurs because the base changes each time the beginning value (revenue or profits, for example) grows. For example, a 10 percent growth in a $1,000 starting point equals $100 and the new base becomes $1,100. If the following year's growth remains at $100, it represents only 9.1% ($100 ÷ $1,100).

3. *The trend is only as reliable as the numbers.* If the numbers used as the base for trends are correct, then the trend is a reliable summary of what is taking place. However, if the numbers are wrong, so is the trend. This is one reason that manip-

ulation of the financial results is so destructive. If revenue is booked a year early or if expenses are deferred a year, the trend is not at all accurate.

4. *Trends can be expressed in many ways, including annual percentage of change or moving averages.* The right way to express a trend depends on what is being reported. Charts showing annual *change* in dollar values are reliable because they are not distorted by the statistical oddity of consistent dollar values and declining percentage growth. The formula for percentage change from year to year is commonly used. It is:

Formula: Percentage Change

$$(N - O) \div O = P$$

where: N = new base value
O = old base value
P = percentage change

On a spreadsheet, cell values are:
A1   N
B1   O
C1   =SUM((A1-B1)/B1)

Example: Microsoft revenues in 2017 were $89,950 (in millions); in 2016, they were $85,320. Applying the formula:

$$(\$89,950 - \$85,320) \div \$85,320 = 5.4\%$$

This change was negative because revenues fell. When revenues rise, the change is positive. However, an analysis of 10 years' change in revenue and profits is not as revealing using this formula, because it is based on year-to-year change without regard to the full decade. Microsoft's results for a full decade are summarized in Table 6.3.

Table 6.3: Revenue and net profit percentage changes
Microsoft Corporation
Annual Percentage change
10 Years ending 2017

Year	Revenue	% change	Net Profit	% change
2017	$89,950	5.4	$21,204	26.2%
2016	85,320	− 8.8	16,798	37.8
2015	93,580	7.8	12,193	−43.8
2014	86,833	11.5	22,074	9.7
2013	77,849	5.6	21,863	28.8
2012	73,723	5.4	16,978	−26.7
2011	69,943	11.9	23,150	23.4
2010	62,484	6.9	18,760	28.8
2009	58,437	− 3.3	14,569	−17.6
2008	60,420	—	17,681	—

These results demonstrate the flaw in using percentage change for a series of years. The outcome does not accurately summarize the trend. For example, annual changes in the reported net profit range from -43.8% to +37.8% This makes development of a real trend difficult to spot. An alternative is to create a visual summary of the history and track the *trend* over many years. In the case of the Microsoft revenue trends, a 10-year bar graph is more revealing than percentage change. Figure 6.1 summarizes net profit on a bar graph. This abbreviated but highly visual reflection shows at a glance how the trend has moved over a full decade. In a report that includes a 10-year summary, the actual dollar values probably need to be included; but the chart tells much more.

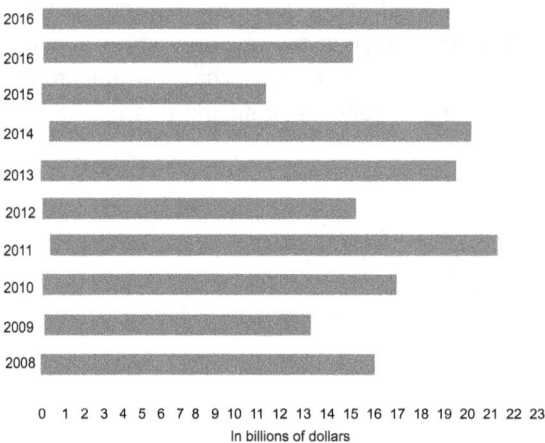

Figure 6.1: Microsoft net profits

Cost of Goods Sold Relationships

The relationship between revenue and net profits is the most important ratio on the income statement. However, when you see that the trend is changing over time, a more detailed analysis of changes is required. You need to evaluate the trend in the cost of goods sold as well as the trend in general expenses. The cost of goods sold is distinct and separate from general expenses. These "costs" are directly attributed to levels of sales volume, whereas expenses occur regardless of increases or decreases in sales.

Costs include:
1. *Merchandise purchased.* The most significant cost is merchandise. For example, a retail organization purchases merchandise at wholesale, marks it up and sells it at retail. The cost of merchandise as reported on the income statement must be adjusted to allow for changes in inventory levels. If a company allows its inventory level to rise by $100,000 during the year, the true cost of goods *sold* must be adjusted. To accurately arrive at the cost of merchandise, the formula is:

Formula: Cost of Merchandise

$$B + M - E = C$$

where: B = beginning balance of inventory
M = merchandise purchased
E = ending balance of inventory
C = cost of merchandise

Example: A company began the year with $1,855,000 in inventory valued at cost. During the year, $7,400,300 in merchandise was purchased. At the end of the year, the inventory was valued at $1,940,000. The cost of merchandise for the year was:

$$\$1,855,000 + \$7,400,300 - \$1,940,000 = \$7,315,300$$

The net cost of merchandise was lower than purchases because inventory was higher at the end of the year. If inventory levels were lower at year-end than at the beginning of the year, the cost of merchandise would be higher than the year's purchases.

The formula on a spreadsheet is:
A1 B
A2 M
A3 E
A4 =SUM(A1+A2-A3)

2. *Direct labor.* Another important cost factor is direct labor. By "direct," this means the payroll costs required to generate product. For example, employees working on an assembly line create the product, so the cost of their labor (and payroll taxes, insurance and other benefits) is part of direct labor. In comparison, payroll for the accounting department and mail room would properly be classified as a general expense. Those expenses do not directly rise or fall with changes in sales volume.

3. *Transportation costs.* Another important direct cost involves moving goods from place to place. In any organization that brings product into a site and then moves it out to another site, transportation costs are going to be incurred; and the higher the sales volume, the higher the transportation costs. For example, a company may import unassembled products from overseas, assemble it in its own plant and then ship it to stores. The cost of bringing products in and then shipping them out are direct costs.

4. *Other direct costs.* Depending on the kind of organization, numerous other direct costs are going to be incurred. In a manufacturing environment, direct costs can be quite elaborate and may involve detailed breakdowns of plant costs. Under the rules of "cost accounting," direct cost analysis and calculation includes the assignment of millions of dollars to different product lines and affects how those products are priced.

In manufacturing, inventory is sub-divided into categories, including raw materials, work in progress, and finished goods. The merchandise and transportation costs may also be assigned to numerous product lines. In companies dealing with a less elaborate series of products, the cost accounting is much easier to calculate. In either case, the important point is to keep a clear distinction between *direct costs* and *general expenses*. The accurate division of these areas of the income statement is required to properly analyze and spot trends.

As direct costs are studied over a period, trends are going to evolve. If the product mix changes, then the relationship between direct costs and revenues will change as well. However, if the mix of products remains unchanged, then in theory, the direct cost relationship should remain steady as well. Any changes should be analyzed and explained. For example, direct labor and transportation costs might rise, without a corresponding increase in pricing. That will reduce profits as direct costs rise as a percentage of revenue. Companies normally pass on price increases to customers by marking up prices to absorb higher direct costs.

The difference between revenue and direct costs is called gross profit. The formula was shown earlier in the chapter. For example, a company reported $13,419,800 in revenue for the year, and direct costs of $7,315,300. Gross profit was:

$$\$13,419,800 - \$7,315,300 = \$6,104,500$$

In tracking gross profit, it is easier to spot changes in the trend when based on a percentage rather than on dollar values. For example, here is a five-year summary of revenue, costs and gross profit using dollar values along:

Year	Revenue	Direct costs	Gross profit
1	$8,005,100	$3,901,600	$4,103,500
2	8,616,300	4,478,500	4,137,800
3	10,007,100	5,371,800	4,635,300
4	11,451,200	6,260,200	5,191,000
5	13,419,800	7,515,300	5,904,500

From this dollars-only summary, it is difficult to determine whether the gross profit trend is moving in a positive or negative direction. The desirable outcome is to maintain the relationship at the same level each year, even when revenue is rising. For this you need to calculate the gross margin. The formula:

Formula: Gross Margin

$$G \div R = M$$

where: G = gross profit
R = revenue
M = gross margin

On a spreadsheet, gross margin is:
A1 G
B1 R
C1 =SUM(A1/B1)

Example: Referring to the previous chart, calculating gross margin results in the following outcome:

Year	Revenue	Direct costs	Gross profit	Gross margin
1	$8,005,100	$3,901,600	$4,103,500	51.3%
2	8,616,300	4,478,500	4,137,800	48.0
3	10,007,100	5,371,800	4,635,300	46.3
4	11,451,200	6,260,200	5,191,000	45.3
5	13,419,800	7,515,300	5,904,500	44.0

The formula applied to the fifth year is:

$$\$5,904,500 \div \$13,419,800 = 44.0\%$$

This version of the analysis reveals what the numbers alone do not: The trend shows that gross margin is eroding each year. This affects profits drastically. For example, if the fifth year's gross margin had been identical to the first year's 51.3%, gross profits would have been:

$$\$13,419,800 \times 51.3\% = \$6,884,357$$

This made a substantial difference in the profitability of the company. Comparing the above to actual gross profits:

$$\$6,884,357 - \$5,904,500 = \$979,857$$

Based on the volume in this example, the change of 7.3% in gross margin cost the company nearly $1 million in net profits. If general expense levels remained steady throughout the five-year period, this is a significant reduction in potential profits. Therefore, the gross margin must be monitored each year, and when negative trends are spotted, why their causes need to be examined. It could be a matter of revising product mark-up for increased costs or needing to improve internal controls due to lower efficiency with higher sales volume.

Revenue and Profitability Trends

The gross margin can affect bottom-line profits over time. The erosion in this margin can be caused by many thing, including rising costs, poor internal controls, or a change in the product mix. However, when it comes to an analysis of expenses, the primary cause for increases is lack of internal control or follow-through on budgets.

Expenses are not directly related to sales. This means that the expenses will occur no matter what. If sales fall from one year to the next, the long-term lease on the company's offices and warehouses continues at the same rate. If sales rise, the level of administrative salaries and wages should not rise dollar for dollar with sales. This is not entirely true, of course. With higher volume, you expect expenses to climb to a degree. So realistically, many general expenses such as salaries must rise as sales revenue rises. In time, the company needs more space, more people, and more locations. The relationship between revenue, costs and expenses should reflect a tracking between sales and costs; and a less reactive rise in general expenses. This relationship is summarized in Figure 6.2.

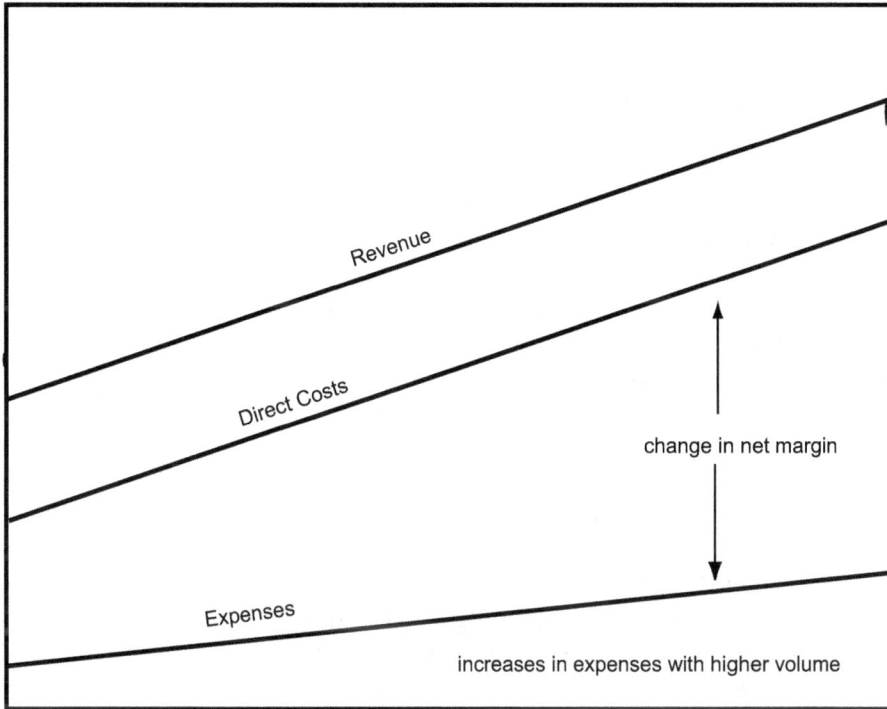

Figure 6.2: Net profit trends

Note how the lines move. You expect costs to follow sales revenue because they are direct; but you also expect expenses to climb only because of higher expense levels and not directly in reaction to revenue changes.

Expenses are broken down into two broad classifications. "Sales expenses" include the travel expenses of marketing and sales personnel, marketing department telephone and postage expenses, and similar kinds of sales-related items. However, sales expenses will rise as sales rise but not at the same rate as costs. The better known "general and administrative" expenses, or "g&a" or simply "overhead," are fixed. You cannot expect administrative salaries, payroll taxes, rent, telephone, and insurance to change just because sales rise and fall, so the "fixed" nature of g&a is its primary attribute. Of course, g&a is going to rise as marketing activity increases, but proper internal controls should help to keep such expenses in line. In fact, budgetary control over items such as office supplies, telephone, and postage are where managers can exert the greatest control and protect the bottom line net profit. One of the greatest dangers in this area is allowing overhead expenses to creep upward when profits are on the rise as well. Remember, however, that the tracking of net return—like gross margin—occurs on a percentage basis. The ideal trend in net profits contains several attributes:

1. *The dollar amount of expenses does not rise as quickly as revenue or costs.* The purpose of separating direct costs from expenses is to track the changes from year to year. You need to ensure that direct costs remain steady and do not creep up as sales revenue rises. You also need to enact internal controls to percentage of expenses from moving upward more than necessary.

2. *Expenses rise more slowly, in response to greater overhead demands and not in response to higher sales volume.* Expenses, by definition, should produce greater efficiency per revenue dollar as revenue levels rise. All too often, you see the opposite in times of higher sales. If expenses rise as a percentage of sales, then the net return is also going to decline.

3. *While the dollar amount of expenses rises, net profit should rise more.* The net return percentage should remain consistent from year to year and should not fall. If it rises, that is a sign of efficiency gained when revenue levels rise; but the net return should always remain the same as a minimum. If expenses outpace revenue, that is a red flag. It means that management is not exerting control over expenses levels. The negative trend in expenses has an eroding effect on long-term profits. For this reason, organizations that end up going out of business often do so after a period of ever-higher sales. They do not control cash flow or expenses. Successful companies know how important it is to hold expense levels in check, *especially* when sales are moving upward rapidly.

The definitions of "profit" and "cash flow" are considerably different. By comparing reported net profits to cash flow (cash coming in and cash going out, these differences become apparent. The most important non-cash expense is depreciation, and any calculation of cash-based returns must be adjusted to remove depreciation from the net profit. A company that invests in capital assets can write off a portion of its original cost each year. As this occurs, the asset value is reduced (through a negative asset account called "accumulated depreciation") and each year's income statement includes an adjustment, made by journal entry, to record depreciation expense. The calculation is mysterious to many, but upon examination, calculating depreciation is not difficult. The next chapter explains.

Chapter 7
Depreciation Calculations

Depreciation is the periodic write-off of capital assets. According to the federal rules:

> Depreciation is the annual deduction that allows you to recover the cost or other basis of your business or investment property over a certain number of years. Depreciation starts when you first use the property in your business or for the production of income. It ends when you either take the property out of service, deduct all your depreciable cost or basis, or no longer use the property in your business or for the production of income.[1]

When a business invests in long-term, or fixed assets (defined as assets with value exceeding one tax year and used in a trade or business), they are placed on the balance sheet as assets. Each year, a portion is transferred to the income statement as expense with an offsetting entry to an account called Accumulated Depreciation. When the asset has been fully depreciated, each year's expense was recorded properly, and the depreciable value of the asset has been reduced to zero (because the sum of accumulated depreciation reduces the asset value).

The accumulated depreciation account is a reduction of the asset. The journal entry to record depreciation is:

Account	Debit	Credit
Depreciation Expense	xxx	
Accumulated Depreciation		xxx

On the balance sheet, the net value of capital assets is the original basis, minus accumulated depreciation. For example, the capital asset section of your company will detail the classes of capital assets and then deduct accumulated depreciation:

Automobiles and Trucks	$ 182,455
Plant Machinery	447,313
Real Estate	2,450,600
Small Tools	84,600
Sub-total	$3,164,968
Less: Accumulated Depreciation	994,260
Net Long-Term Assets	$2,170,708

---

1 www.irs.gov, form 4562

DOI 10.1515/9781547400638-007

Basic Depreciation Rules

The rules for calculating and deducting depreciation are set by the Internal Revenue Service. To get an overview of the current rules, go to the website *www.irs.gov* and download the publication "Depreciation and Amortization" (Publication 4562). Important rules to keep in mind:

4. *The recovery period is determined by the type of asset.* The IRS has developed a series of specific recovery periods based on the nature of the asset involved. For example, assets like automobiles and computers are likely to lose value and need replacement much faster than real estate; as a result, the differences in recovery periods are substantial.

5. *The method used to calculate depreciation is determined by the recovery period.* Also prescribed by the IRS are the kinds of depreciation you can use. For some short-life assets, you can accelerate the method so that more write-off takes place at first, and less later when the asset is worth much less. Long-term assets, notably real estate, cannot be depreciated on an accelerated basis, but can be written off only using the "straight-line" method (in which the same amount is written off each year).

6. *In the case of real estate, only improvements (buildings, etc.) can be depreciated. Land cannot be depreciated.* A special case arises for depreciation of real estate. You can never depreciate land, which must be kept on the books indefinitely at cost, to be removed only when the property is sold.

7. *The basis for depreciation is always the net cost plus improvements, and never the current market value.* An oddity in the accounting system is that assets must be depreciated based on actual cost, plus improvements if any. You can never increase the book value of real estate. For example, a company purchases its land and building for $1.6 million. Ten years later its market value has increased to more than $5 million. Even so, the value reported on the balance sheet *declines* each year as the building's depreciation is written off. The increased market value is not acknowledged on the balance sheet.

8. *Depreciation begins not when an asset is purchased or paid for, but when it is placed into service.* You might purchase an asset this year but not begin using it until next year. In this situation, you cannot begin to claim depreciation expense right away. Depreciation begins only in the year the asset is placed into service.

9. *Some elections can be made to alter the method used for depreciation; however, that election must be applied to all assets in the same class, placed into service in the same year.* A variety of depreciation-based elections can be made each year. However, you cannot pick and choose. Each asset in the year's recovery period must be subjected to the same elections each year. For example, in some recovery periods you can claim accelerated depreciation; you can elect to use the straight-line method for assets placed into service that year.

Straight-line and Declining Balance Depreciation

Two kinds of depreciation are used for virtually all calculations: straight-line and declining balance. Straight-line is claiming the same amount each year over the life of the asset (its recovery period). Declining balance is a calculation allowing you to claim more depreciation in the early years and less in the later years. Under the IRS-published calculations, assets that can be accelerated use the declining balance method and then reverts to straight-line in later years.

Straight-line depreciation is widely favored in most of the world, and for good reason:

> Straight-line depreciation (SL) appears to be a crude procedure that is unsupported by economic logic. Nevertheless, internationally, it is the most widely used method of allocating the costs of fixed assets to accounting periods by way of depreciation charges. Many authors attribute its use to its simplicity.[2]

The simplicity of straight-line certainly is one of its appeals. However, it is not the entire story. The ideal form of depreciation should accurately reflect the declining true value of capital assets, and this is not necessarily accomplished through straight-line depreciation. Many assets (such as autos and trucks) lose more value in the earlier years than later. Other assets, such as improvements to real estate, tend to appreciate over time even while depreciation reduces book value every year.

Even with its many flaws in accurately reflecting the value of an asset as it ages, straight-line depreciation is appealing because it is easily understood. However, complicating the use of any depreciation method is the fact that it is subject to certain rules imposed by the IRS. For example, to calculate straight-line depreciation, you need to pay attention to the rules governing first-year levels you can take. This is explained later in this chapter. The basic equation for straight-line depreciation involves a single step. The formula:

Formula: Annual Straight-line Depreciation

$$B \div Y = D$$

where: B = basis
Y = years in the recovery period
D = annual depreciation

---

2 Green, C. D.; Grinyer, J.R. & Michaelson, R. (December 2002). A Possible Economic Rationale for Straight-Line Depreciation. *Abacus*, Volume 38, No. 1.

On a spreadsheet, cells are:

A1   B
B1   Y
C1   =SUM(A1/B1)

Example: You purchased an asset for $7,500 and placed it into service on January 1 this year. Using the straight-line method and assuming the asset qualifies for a five-year recovery period, the formula is:

$$\$7,500 \div 5 = \$1,500 \text{ per year}$$

To calculate the same level of depreciation but monthly, the formula is:

Formula: Monthly Straight-line Depreciation

$$B \div (Y \times 12) = D$$

where: B = basis
Y = years in the recovery period
D = monthly depreciation

On a spreadsheet:

A1   B
B1   =SUM(Y*12)
C1   =SUM(A1/B1)

Using the same example as before, monthly depreciation is:

$$\$7,500 \div (5 \times 12) = \$125 \text{ per month}$$

The method for calculating declining balance depreciation is more complex. Under this method, you can claim more than the straight-line amount per year in the early years, and less as time moves forward. There are two common methods: 150% and 200%. Under the 150% method, you can deduct 150% of the straight-line depreciation in the first year. Subsequent years are calculated as 150% of the balance remaining. You begin with the original basis, deduct the first year's depreciation, and then repeat the calculation. The formula for 150% declining balance depreciation:

Formula: Declining Balance (150%) Depreciation

$$((B - P) \div Y) \times 150\% = D$$

where: B = basis
P = previous years' accumulated depreciation
Y = years in the recovery period
D = annual depreciation

Under the accumulated depreciation method, the basis for depreciation is recalculated each year. The basis declines by the total of previously claimed depreciation.

A five-year summary reveals the following fields on the spreadsheet:

	A	B	C
1	7,500	2,250	5,250
2	5,250	1,575	3,675
3	3,675	1,103	2,573
4	2,572	772	1,800
5	1,800	540	1,260

On a spreadsheet:

A1	B
B1	=SUM((A1/5)*1.5)
C1	=SUM((A1-B1)
A2	=C1
B2, C2	copy B1 and C1; paste to B2 and C2
row 3	copy all cells, row 2; paste to row 3

Example: You purchased an asset worth $7,500 and you want to apply the 150% declining balance formula. For the first year, this is:

$$(($7,500 - $0) \div 5) \times 150\% = $2,250$$

For the second year, the basis is reduced by the amount of depreciation previously claimed:

$$(($7,500 - $2,250) \div 5) \times 150\% = $1,575$$

For the second declining balance, the same formula is used but the multiplier of 1.5 (150%) is replaced with a multiplier of 2 (200%):

Formula: Declining Balance (200%) Depreciation

$$((B - P) \div Y) \times 200\% = D$$

where: B = basis
P = previous years' accumulated depreciation
Y = years in the recovery period
D = annual depreciation

The spreadsheet entries:

A1   B
B1   =SUM((A1/5)*2)
C1   =SUM((A1-B1)
A2   =C1
B2, C2   copy B1 and C1; paste to B2 and C2
row 3   copy all cells, row 2; paste to row 3

The outcome over five years:

	A	B	C
1	7,500	3,000	4,500
2	4,500	1,800	2,700
3	2,700	1,080	1,620
4	1,620	648	972
5	972	389	583

Because the schedule in these examples calls for a five-year depreciation recovery period, the balance at the end of the fifth year is normally written off in the sixth year. In the tables published by the IRS (shown later in this chapter), declining balance is applied for a specific number of years, and then the schedule reverts to the straight-line method.

Class Lives and Recovery Periods

The rules for depreciating assets include a series of class lives, each containing a recovery period. When assets do not fit into the narrowly defined recovery periods, they are depreciated using the closest one to the reasonable useful life of the asset. The properties classified by recovery period are listed in the IRS instructions for depreciation.

The IRS also specifies several *conventions* for calculating first-year depreciation. Depending on when property was purchased and placed into service, one of the three conventions will apply. These are:
1. *Half-year convention.* This system applies to most short-term properties but not to real estate. The assumption under this convention is that no matter when property was placed into service, first-year depreciation is one-half of the full-year calculated rate. For example, under this convention, the first year's depreciation was calculated at $2,250. Under the half-year convention, the deduction in the first year would be $1,125.

Formula: Half-year Convention

$$C \div 2 = D$$

> where: C = calculated full-year depreciation
> D = depreciation, first year

On a spreadsheet:
A1 C
B1 =(SUM(A1/2))

2. *Mid-quarter convention.* This method is used in some circumstances and for some kinds of property (see instructions for depreciation published by the IRS). The assumption is that property was placed into service exactly mid-way in the quarter. The mid-point in any one quarter is equal to 1.5 months, so the calculation varies based on the quarter.

    Tables summarizing annual depreciation allowed using the half-year convention and based on 150% declining balance, are shown in Table 7.1.

Table 7.1: Depreciation—150% declining balance with half-year convention

Year	RECOVERY PERIOD					
	5-year	7-year	10-year	12-year	15-year	20-year
1	15.00%	10.71%	7.50%	6.25%	5.00%	3.750%
2	25.50	19.13	13.88	11.72	9.50	7.219
3	17.85	15.03	11.79	10.25	8.55	6.677
4	16.66	12.25	10.02	8.97	7.70	6.177
5	16.66	12.25	8.74	7.85	6.93	5.713
6	8.33	12.25	8.74	7.33	6.23	5.285
7		12.25	8.74	7.33	5.90	4.888
8		6.13	8.74	7.33	5.90	4.522
9			8.74	7.33	5.91	4.462
10			8.74	7.33	5.90	4.461
11			4.37	7.32	5.91	4.462
12				7.33	5.90	4.461
13				3.66	5.91	4.462
14					5.90	4.461
15					5.91	4.462
16					2.95	4.461
17						4.462
18						4.461
19						4.462
20						4.461
21						2.231

A table using 200% declining balance and based on the half-year convention is shown in Table 7.2.

Table 7.2: Depreciation—200% declining balance with half-year convention

Year	RECOVERY PERIOD			
	3-year	5-year	7-year	10-year
1	33.33%	20.00%	14.29%	10.00%
2	44.45	32.00	24.49	18.00
3	14.81	19.20	17.49	14.40
4	7.41	11.52	12.49	11.52
5		11.52	8.93	9.22
6		5.76	8.92	7.37
7			8.93	6.55
8			4.46	6.55
9				6.56
10				6.55
11				3.28

Formula: Mid-quarter Convention

1st quarter

$$C ((1.5 \div 12) \times 7) = D$$

2nd quarter

$$C ((1.5 \div 12) \times 5) = D$$

3rd quarter

$$C ((1.5 \div 12) \times 3) = D$$

4th quarter

$$C ((1.5 \div 12) \times 1) = D$$

where: C = calculated depreciation
D = first-year depreciation

Example: The calculated depreciation for the first year is $2,250 and property was placed into service during the second quarter. Using the mid-quarter convention, first-year depreciation is:

$$\$2,250 ((1.5 \div 12) \times 5) = \$1,406$$

The spreadsheet entries for the mid-quarter convention are:

A1 C
B1 =SUM(1.5/12)*x
C1 =SUM(A1*B1)

In this calculation, the value of 'x' is the factor for each quarter:

1st quarter   7
2nd quarter   5
3rd quarter   3
4th quarter   1

3. *Mid-month convention.* For depreciation on real estate, the assumption is made that property was placed in service half-way through the applicable month. Because there are 24 half-months in each year, the formula is based on calculated half-way points using fractions of 24ths. The formula:

Formula: Mid-month Convention

$$C\,(n \div 24) = D$$

where: C = calculated depreciation

n = monthly fraction

D = first-year depreciation

The monthly fraction is based on 24 half-months. For example, in the first month of the year, 'n' is valued at 23 using the half-month calculation. Depreciation is going to be equal to 23/24ths of the full-year calculated value. The value of 'n' is equal to:

Month	n	Month	n
1	23	7	11
2	21	8	9
3	19	9	7
4	17	10	5
5	15	11	3
6	13	12	1

Spread cell entries:

A1 C

B1 =SUM(n/24)

C1 =SUM(A1*B1)

Example: You purchased real estate last year and placed it into service during the third month. The calculated depreciation per year is $8,100. Using the mid-month convention, the formula is:

$$\$8,100\,(13 \div 24) = \$4,387$$

Residential real estate can be depreciated using only the straight-line method, and the period for depreciation is 27.5 years. A summary of each year's percentage of depreciation allowed is summarized in Table 7.3.

Table 7.3: Depreciation—residential real estate with mid-month convention

Month	The month in the first recovery period the property was placed in service:				
	Year 1	Years 2 to 9	Years 10, 12, 14, 16, 18, 20, 22, 24, 26	Years 11, 13, 15, 17, 19, 21, 23, 25, 27	Year 28
1	3.485%	3.656%	3.637%	3.636%	1.970%
2	3.182	3.636	3.637	3.637	2.273
3	2.879	3.636	3.637	3.637	2.576
4	2.576%	3.636%	3.637%	3.637%	2.879%
5	2.273	3.636	3.637	3.637	3.182
6	1.970	3.636	3.637	3.637	3.485
7	1.667%	3.636%	3.636%	3.637%	3.788%
8	1.364	3.636	3.636	3.637	4.091
9	1.061	3.636	3.636	3.637	4.394
10	0.758%	3.636%	3.636	3.637%	4.697%
11	0.455	3.636	3.636	3.637	5.000
12	0.152	3.636	3.636	3.637	5.303

Non-residential real estate is depreciated using the straight-line method, over 39 years. A summary of each year's depreciation is shown in Table 7.4. Note that a 40th year's depreciation is required due to partial first-year allowance.

Table 7.4: Depreciation—non-residential real estate with mid-month convention

Month	Month in the first recovery year property is placed in service:		
	Year 1	Years 2 to 39	Year 40
1	2.461%	2.564%	0.107%
2	2.247	2.564	0.321
3	2.033	2.564	0.535
4	1.819%	2.564%	0.749%
5	1.605	2.564	0.963
6	1.391	2.564	1.177
7	1.177%	2.564%	1.391%
8	0.963	2.564	1.605
9	0.749	2.564	1.819
10	0.535%	2.564%	2.033%
11	0.321	2.564	2.247
12	0.107	2.564	2.461

Depreciation Calculations for Real Estate

The depreciable basis of some assets is not always apparent. You may need to prorate the total basis. Proration—the division of a dollar value between two or more categories—is commonly used in accounting and other applications. For example, a company-wide utility expense is likely to be prorated to each department based on floor space, numbers of employees, or other logical proration methods.

In real estate, closing statements prorate property taxes, insurance, interest, and utilities between the buyer and seller. This is broken down based on days. For example, a property tax bill is paid twice per year. A transaction closes on the 43rd day of the current tax period. The seller is prorated 43/180ths of the bill and the remaining 137/180ths is assigned to the buyer.

The same treatment of a purchase price must be used to prorate the cost of property between land (which cannot be depreciated) and improvements (the building, which can be depreciated). However, a sale is not normally detailed between these separate values. For this reason, at least three methods can be used to accurately prorate the total basis between land and improvements:

1. *Insurance.* The property insurance is based on estimated replacement value of the improvements. Since this estimate is developed through a study of local costs, it is likely to be an accurate cost estimate for the depreciable portion of the investment.

2. *Appraisal.* The appraisal undertaken for a lender will break down the estimate of market value between land and improvements, based on recent sales of similar properties in the same area.

3. *Assessed value.* The local taxing authority bases its tax levy on its own estimate of land and improvement value. Although this often is far below market value, the division between the two elements is a reliable base for dividing the actual cost.

Proration of total cost is based on the use of some reliable and applicable base. In the case of accounting decisions, expenses may be divided based on square footage of a department or the number of employees, for example. The closing of a real estate transaction relies on the number of days before and after the closing date to divide up property taxes, utilities, rent, and other shared items. The division of real estate cost between land and improvements can be based on the values used for insurance coverage, appraisal, or local assessment. The formula:

Formula: Proration

$$V (a / (a + b)) = P_a$$

$$V (b / (a + b)) = P_b$$

where: V = value to be prorated
a = proration base factor a
b = proration base factor b
P_a = prorated value of a
P_b = prorated value of b

In this two-part formula, the values 'a' and 'b' add up to the total of the base; and the values of 'P_a' and 'P_b' equal the total of 'V.' On a spreadsheet:

A1 V
B1 a
C1 b
D1 =SUM(a/(a+b))
E1 =SUM(b/(a+b))
F1 =SUM(A1*D1)
G1 =SUM(A1*E1)

Example: Your company purchased its land and building for $2,650,000. To set up a depreciation schedule, you need to isolate the values of land and improvements. Land cannot be depreciated, so this calculation relies on proration. The assessed value of the property as of the most recent assessment and tax statement was:

Land	$ 250,000
Improvements	1,665,000
Total	$1,915,000

Using this breakdown as the base, you assign land the value 'a' and improvements the value 'b.' The purchase price of $2,650,000 is the 'V' you need to break down. Applying the formula and rounding the outcome to the closest $100:

$2,650,000 (($250,000 ÷ ($250,000 + $1,665,000) = $346,000
$2,650,000 (($1,665,000 ÷ ($250,000 + $1,665,000) = $2,304,000

Now that the values of land and improvements are separated, you can apply depreciation correctly. In this example, $2,304,000 can be depreciated. Non-residential property is depreciated using the straight-line method over 39 years. The prorated value of land, or $346,000, remains as an asset on the balance sheet and never changes. It can be removed only when the property is sold.

This proration formula can be used in other applications. For example, when a single value (V) must be divided among multiple people, departments or recipients, the sum of all base values is added up and expressed as fractions of the total. These are then applied to the single "V" to get the prorated outcomes.

The accuracy of the formula is proven by adding together the prorated values and ensuring that they equal the starting value:

Formula: Proof of Proration

$$P_a + P_b = V$$

where: P_a = prorated value of a
P_b = prorated value of b
V = value to be prorated

On a spreadsheet program:
F1 P_a
G1 P_b
H1 =SUM(F1+G1)

In the preceding example:

$$\$346,000 + \$2,304,000 = \$2,650,000$$

This proves that the formula was applied correctly.

Home Office Depreciation

Those who work from a home office can calculate depreciation and other expenses, and this represents another case where proration must be used. In fact, to calculate depreciation on a qualified home office, two proration steps are necessary. To qualify for a home office deduction, you need to meet a two-part test: First, you must use the office on a "regular and exclusive" basis. This means the space cannot double for other purposes. And second, it must be your principal place of business; in other words, if you conduct business elsewhere, you need to be able to prove that the primary place of business is the home office.

Proration is not complex. It means that a total is divided into parts based on a logical base (for prorated expenses, number of days is used, and for home office deduction, it is based on square feet). The intention is to accurately calculate how much of a total is allowed as a deduction based on the business use of a home office. For example, if your home is 2,000 square feet and your office is used exclusively for business and measures 200 square feet, you are allowed a deduction of 10% of total depreciation calculated for the building.

If you are a manager employed by someone else, but you also work at home, you still need to meet these criteria. For complete information, go to the IRS website at www.irs.gov and download the free publication 587, "Business Use of Your Home." Deductions are reported on form 8829, "Expenses for Business Use of Your Home." This can also be downloaded and printed in PDF form from the IRS website.

In calculating the deductions, you are allowed if you qualify, you first need to determine the prorated space in your home that qualifies. If you live in a home with total square feet of 3,200 and you qualify for a home office deduction for one room of 350 square feet, the prorated portion of your home is:

$$350 \div 3{,}200 = 11\% \ (rounded \ up)$$

You can deduct 11% of qualified expenses. This includes utilities, homeowners' insurance, insurance and depreciation. For depreciation, you need to use the proration formula to break down the value of land and improvements, all based not on current value but on your original purchase price. For example, if you purchased your home for $289,000 you may break down the land and improvements value based on your most recent assessor's statement. This shows:

Land	$50,000
Improvements	182,000
Total	$232,000

Applying the proration formula:

$$\$289{,}000 \times (\$50{,}000 \div (\$50{,}000 + \$182{,}000)) = \$62{,}300$$

$$\$289{,}000 \times (\$182{,}000 \div (\$50{,}000 + \$182{,}000)) = \$226{,}700$$

This conclusion shows that you can deduct home office expenses based on improvement value of $226,700. Residential real estate can be depreciated over 27.5 years; in addition, the proration of floor space revealed that the home office represents only 11% of the total. The formula for proration of home depreciation.

Applying this to the example:

$$((\$289{,}000 \times (\$182{,}000 \div \$232{,}000)) \div 27.5) \times (350 \div 3{,}200) = \$902.00$$

Formula: Home Office Depreciation

$$((B \times (I \div A)) \div 27.5) \times (of \div tf) = D$$

where: B = basis (purchase price)
I = improvement value (assessed)
A = assessed value, total
of = office square feet
tf = total square feet
D = depreciation allowed

On a spreadsheet:
A1 B
A2 I
A3 A
B1 of
B2 tf
C1 =SUM((A1*(A2/A3)))
C2 =SUM(C1/27.5) (note: 27.5 is the number of years depreciation can be claimed)
C3 =SUM((C2*(B1/B2)))

This demonstrates that, based on the size of the home office and using the breakdown of the assessed value:
1. A total of 78% of total cost is depreciable ($182,000 ÷ $232,000, applied to pur-chase price of $289,000).
2. Only 11% of total floor space can be depreciated as an office in the home.
3. A qualified home office deduction is $902 per year.

Amortization

In addition to depreciation, some kinds of expenses are set up as assets and amor-tized over a period of years. In most instances, amortization is calculated using the straight-line basis. Like depreciation, the annual expense is booked to reflect assign-ing the cost to several years. However, rather than setting up a negative asset account, each year's expenses is usually deducted from the asset value on the balance sheet.

Amortization is applied to general expenses that are paid in advance but applied over several years. For example, your company paid a 36-month insurance premium in March. This expense applied over a three-year period so recording the entire expense in the current year would not be accurate. Instead the payment is set up as a "prepaid asset" on the balance sheet. If the total paid was $3,852, the entry is:

Account	Debit	Credit
Prepaid Assets	$3,852	
Cash		$3,852

The proper amount to be expensed is equal to 1/36th per month. In the first year, this period covers March through December, or 10 months. Amortization will take place based on this breakdown. The formula for determining monthly amortization is:

Formula: Amortization

$$C \div M = A$$

where: C = total cost
M = months to amortize
A = amortization per month

On a spreadsheet:
    A1   C
    B1   M
    C1   =SUM(A1/B1)

Referring to the previous example, the current year's expense is 10/36ths of the total:

$$\$3,852 \times (10 \div 36) = \$1,070$$

The second and third years will each equal 12/36ths (one-third) of the total:

$$\$3,852 \times (12 \div 36) = \$1,284$$

The fourth year will recognize the remaining two months of the total:

$$\$3,852 \times (2 \div 36) = \$214$$

At the end of the 36 months, the prepaid asset value will be zero.

A second type of amortization is not intended to properly recognize general expenses, but to write off certain kinds of costs that apply over many years. This is more like depreciation of capital assets and is applied to tangible costs like research and development or the cost of acquiring a lease. In these instances, the cost is amortized over the applicable period determined by accounting rules. Amortization can also be applied to intangible assets, those items shown on the balance sheet but containing no physical value. These include goodwill, going concern value, brand value, or covenants not to compete. These valuations often are determined at the time of a merger or acquisition and booked according to how a contract is worded. In these instances, the amortization booked as expense each year reduces the value of the intangible asset until it reaches zero.

The next chapter moves into the realm of reports. Within the body of a report, you can simply list financial information without comment, but that is ineffective. You can also explain the numerical information in a narrative section. Or you can apply ratios and shorthand expressions of the numbers to add powerful and convincing arguments about not only the latest set of financial results, but also how they augment or contradict an established trend.

Chapter 8
Bringing Reports to Life: Powerful Arguments with the Numbers

If you were not called a "manager," you might be named a "reporter." Managers are expected to report in various ways, to summarize and convey information, make recommendations, give early alerts to emerging problems, and cut costs. The written reports that you prepare and distribute say a lot to others about your professionalism and your ability to communicate.

Even so, some managers struggle with reports because they have not been provided with a few basic skills. These include the ability to *summarize* financial information, explain what it means, and combine it into an informative narrative. The tendency is to simply *include* financial reports, avoid explanation, and provide no narrative supplements. Whether you are working with a complete set of financial statements, public disclosure, departmental budgets, or a marketing research paper about a new product, you cannot avoid financial data; and once you realize how effective it is to manage the numbers, you will also find that even the dullest topic can spring to life and work as an effective device for conveying your ideas to other managers.

In public reporting of financial data, required under regulatory rules of publicly listed companies, the format and disclosure requirement are strict. Even so, the pressure on executives to meet or exceed earnings expectations causes distorted or even false results in some cases. A comprehensive survey of this problem documented:

> ... a willingness of corporate executives to routinely sacrifice shareholder value to meet earnings expectations or to smooth reported earnings ... Real earnings management, which might include deferring a valuable project or slashing research and development expenditures, is almost always value decreasing.[1]

This fact reflects ongoing practices in both management and accounting sectors, to prefer favorable earnings reports to accuracy. When reviewing reported financial results, you must be aware of the potential inaccuracy of what corporate executives have decided, often with the willing approval of outside auditing firms.

---

1 Graham, J., Harvey, C., & Rajgopal, S. (2006). Value Destruction and Financial Reporting Decisions. *Financial Analysts Journal*, 62(6), 27–39.

DOI 10.1515/9781547400638- 008

Picking Your Report Format

The problem of financial distortion is not immediate if you prepare your own reports and are devoted to accuracy in your conclusions. In this case, your first task is to pick the best report format for the topic at hand. No one format is going to work best for every kind of report; and generally, remember that the longer the report, the less chance it has of even being read. A three-page summary is much more effective than a highly detailed 80-page analysis.

A few ways to keep your report short:

1. *Provide a summary only but have supporting documents ready.* Assume that anyone receiving your report is going to give it a glance only. Whether produced online or in hard copy, the report is rarely going to be read cover to cover. A two- or three-page report that gives your conclusions and highlights is far more effective than a longer report. Those readers who want to see the supporting documentation can be referred to a back-up report or online link. However, most readers will accept your conclusions without asking for proof. It is important to have the supporting documents ready, but just as important to provide them only on request.

2. *Organize the report if your audience will only read the first paragraph.* Put your most important point on the first page. Be clear, concise and brief. Many people, even those who need the information, will only read the first paragraph. What one thing do you need those readers to know? That must go right in the front of the report. A second suggestion is to include your conclusion in the title of the report. A title like "Keys to a 10% Growth in Market Share" is much more interesting than "Third Quarter Marketing Forecast."

3. *Never include pages full of numbers in the body of the report. Charts and graphs say much more.* As important as the dollar values and statistics are, they are boring to review. The more numerical information you include in your report, the less people will read it and, more to the point, the less likely they are to comprehend their meaning. Remember, your report's purpose is to reveal information and to express ideas, not to fill pages with columns and rows of numbers. Refer readers to an appendix if they want the details, or to an online site where your spreadsheets are included with all the details and explanations. But in the report itself, leave out the details and focus on the information you need your reader to hear.

The formatting of a report should be based around these few concepts. The shorter the report, the better. Next, you need to determine whether the report should be organized to provide in-depth narratives or a very brief look at (a) conclusion, (b) recommendations, and (c) details. For example, if your report is a bid for an engineering project, you will need to provide statistical information and explain how you will approach the job. You are competing with others who are also bidding on the same job. If your report is providing a cost-reduction recommendation, you will need to explain the problem and then document your solution. And if your report addresses methods for increasing market share for your products, you are going to need an analysis of customers, the competition, and methods for achieving the market expansion goals.

In other words, the format of the report is dictated by its intention. Even so, the same basic rules apply. Keep it short, avoid massive numerical sections, and prioritize. You can probably think of dozens of reports you might be asked to prepare, or that you will need to review when prepared by other managers or employees. If the same format is applied in each case, the effectiveness will be lost. Even if most people are using a similar format, be ready to think creatively. The key to selecting the right format is based on how you want to convey the essential information your report contains.

The report may consist of both narrative and financial sections, or a combination of both. Even if you do not want to place the details elsewhere, you can assign financial details to an appendix and then explain them in summarized form. Using graphics, ratios, and narratives, you should be able to create a "user friendly" reporting format that works best in each situation. By nature, reports fall into two classifications. The recurring format is in the minority, and the more common one-time report is the rule.

Narrative Sections

A report can consist primarily of narrative sections. By itself, a long narrative is uninteresting and difficult to read, especially for a dry topic. However, if the narrative is broken up with a few short but informative departures, this vastly improves the report. These can include:

1. *Sidebars.* A sidebar is a related thought and a form of emphasis that is set aside from the narrative and boxed off in some way. This is used freely in magazine articles. Also termed a "call-out," the sidebar breaks up the discussion and brings to the reader's attention the primary idea you want to convey. The use of sidebars should be limited; but for example, if there are three major thoughts your reports needs to get across, use sidebars. Even a casual glance at your report will highlight the main ideas.
2. *Tables.* It is very difficult to read a report containing a lengthy explanation of the numbers. A table can summarize a numerical report that must be included. For example, compare the following two paragraphs:

    Approach 1

    During the quarter, several important improvements occurred in the marketing activity of the branch. The western division moved sales volume from $425,000 up to $519,000 over the prior quarter, a growth rate of 22.1%. The southern division saw growth from $318,000 in sales to $341,000, an improvement over the prior quarter of 7.2%. Central division grew in sales volume from $543,000 to $602,000, or 10.9%. Northern division, the newest expansion in the branch, grew from $127,000 to $172,000, an improvement of 35.4%. Only the eastern division had a decline, changing from the prior quarter's total of $288,000 to current quarter of $251,000, a decline of 14.7%.

Approach 2

Four out of five divisions in the branch experienced growth during the quarter, summarized below:

Division	Prior	Current	Change
Western	$425,000	$519,000	22.1%
Southern	318,000	341,000	7.2
Central	543,000	602,000	10.9
Northern	127,000	172,000	35.4
Eastern	288,000	251,000	– 14.7

The decline is attributed in part to the loss of our divisional manager during the quarter. The decline is not expected to continue.

The second approach summarizes the financial information in a separate table and adds a brief explanation for the disappointing results in one out of five divisions. The second approach is preferable because it is easier to digest and to comprehend.

3. *Charts and graphs.* Using charts and graphs is an excellent way to liven up a report. Financial information is difficult for people to understand in context, and it is easier when summarized in visual form. Figure 8.1 shows the same data as the previous example but in bar graph form.

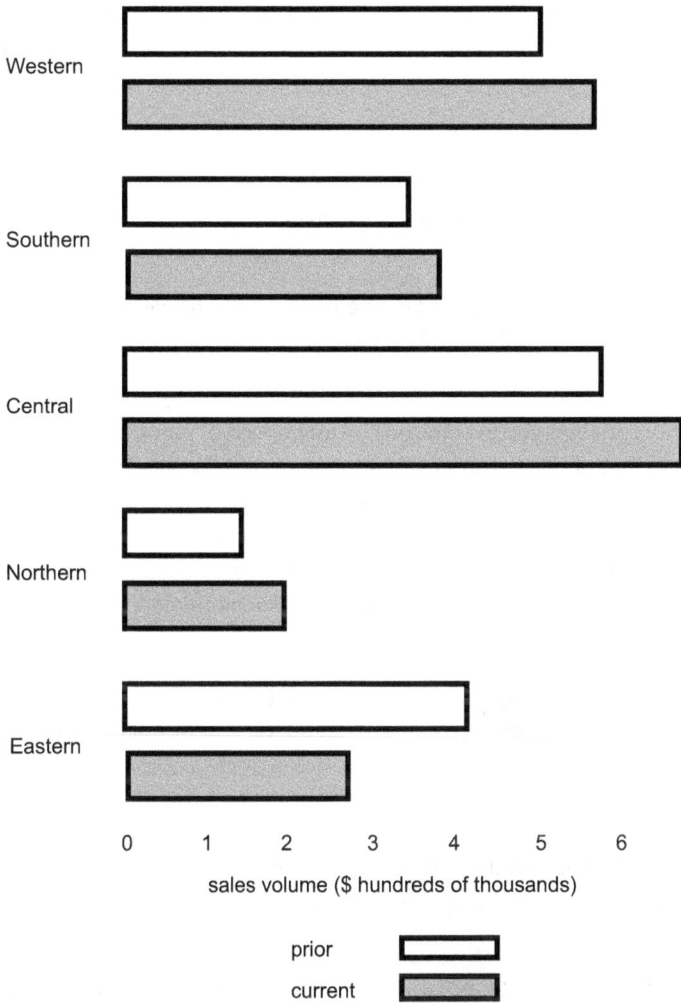

Figure 8.1: Volume of sales by division, quarterly comparison

This is a vast improvement over even the table. This not only shows the quarter-to-quarter change, but also provides a visual comparison between divisions. This method of conveying information is far superior to the narrative paragraph and to the table.

4. *Footnotes.* Using footnotes livens up a report, while enabling you to place secondary thoughts apart from text. It is also a subtle form of emphasis that breaks up an otherwise monotonous report. Returning to the previous example, the discussion of sales volume by branch division could be set up in the following way:

Four out of five divisions in the branch experienced growth during the quarter, summarized below:

(Graph could be inserted here)

The decline in the Eastern Division is not expected to continue.[1]

------

1 The eastern divisional manager left the organization during the quarter, creating temporary turmoil and a leadership gap. We have since replaced him and current trends are eturning to prior levels.

This use of a footnote enables you to elaborate on an important point, but without breaking up the rhythm of the narrative. Footnotes can also be used for non-essential but interesting facts that add to the tone of your report.

5. *Narrative forms of emphasis.* The most apparent form of emphasis within a report is emphasis. It is easily overused, however and should be avoided. Even a very short paragraph loses its impact if overly emphasized. For example, consider this short paragraph:

Marketing volume has been exceptionally strong this quarter due to the recent acquisition of our only major competitor. However, we also expect a finite growth curve as we saturate a limited customer base.

This statement is extremely important. In wanting to ensure that its importance is appreciated, there is a tendency to overuse narrative emphasis. Four examples:

Italicizing
Marketing volume has been exceptionally strong this quarter due to the recent acquisition of our only major competitor. However, we also expect a finite growth curve as we saturate a limited customer base.

Underlining
<u>Marketing volume has been exceptionally strong this quarter due to the recent acquisition of our only major competitor. However, we also expect a finite growth curve as we saturate a limited customer base.</u>

Bold-facing
Marketing volume has been exceptionally strong this quarter due to the recent acquisition of our only major competitor. However, we also expect a finite growth curve as we saturate a limited customer base.

Capitalizing
MARKETING VOLUME HAS BEEN EXCEPTIONALLY STRONG THIS QUARTER DUE TO THE RECENT ACQUISITION OF OUR ONLY MAJOR COMPETITOR. HOWEVER, WE ALSO EXPECT A FINITE GROWTH CURVE AS WE SATURATE A LIMITED CUS-TOOMER BASE.

Larger font

Marketing volume has been exceptionally strong this quarter due to the recent acquisition of our only major competitor. However, we also expect a finite growth curve as we saturate a limited customer base.

All these forms of emphasis are obnoxious and tedious because they are used in excess. With nearly universal use of easy-to-use automated report composition, all these forms of emphasis (as well as others, such as text highlighting, formatting changes, or combined forms of emphasis (such as italics, underlines and all-caps used together) are even more ineffective. Even bad habits like capitalizing every word in a sentence or placing quotation marks around non-quote items are merely distracting. For example:

> "Marketing Volume" Has Been Exceptionally Strong This Quarter Due To The Recent Acquisition Of Our Only "Major Competitor." However, We Also Expect A "Finite" Growth Curve As We Saturate A "Limited" Customer Base.

The distraction in this paragraph makes the message difficult to comprehend. The capitalization and use of quotes adds nothing; in fact, it makes the message less clear.

Financial Sections

Many reports contain distinct sections of narrative material and then financial material. Whether financial reports are provided in an appendix or mixed in with narrative explanations, the use of a large volume of numbers is a problem. Many readers are going to skip the financial rows and columns or fail to grasp their importance, due to the methods used in presentation. Report preparers miss a great opportunity to bring the numbers to life by not considering how to most effectively design the report to maximize its effectiveness.

When you must prepare a report heavily involved with financial data, you face a challenge. Many reports contain a great deal of financial reporting, but the most effective reports are those that are readable, visual and short. Even when the financial sections are lengthy, there are good solutions:

1. *Use an appendix.* Keep the body of your report to as few pages as possible. Put supporting documentation in an appendix so that readers can refer to it without extra steps.

2. *Link to financial data separately.* Place financial supporting data in a separate online link and provide the link within the report at spots where financial data are referenced.

3. *Summarize the key items but exclude the details.* Highlight the important conclusions you reach based on the financial data, but don't automatically assume that your readers want to see those details.

4. *Offer to provide back-up on request but exclude from the report.* The short report is appreciated by most people and is more likely to be read. For those wanting all the details, invite requests. You probably will not get many people asking for the financial back-up but offer to provide it.

Any of these methods or combinations of them make your reports more efficient. As a final thought, consider breaking down full-page financial reports into smaller sections. For example, if part of your report includes a complete set of financial statements, present the assets on a half-page within a narrative section, and then present liabilities and net worth later. You can also break down revenues/costs/gross profit and expenses/net profit into different sections. A widespread assumption is that financial reports must be provided intact, but this makes little sense. The important point is that your reports should be digestible, and a lot of numbers are difficult to manage all at once. Break them up. The same logic applies to numbers in any form: marketing reports or forecasts, budgets, cost breakdowns, manufacturing estimates … all of these are more effectively reporting in sections.

Any time your report includes a full page of numerical information, question how it can be broken down into smaller pieces. No one likes poring over a page full of rows and columns of numbers.; They will do better with a limited presentation, numbers represented in graphs or explained with ratios, and key information emphasized and explained.

Some methods you can use to mathematically summarize the numbers:

1. *Use percentages instead of dollar values for broad statements.* Avoid the use of dollar values in narrative sections. They are difficult to comprehend, especially in comparative form. You can be far more effective using percentage-of-change statements. Some examples:
   - The revenue total rose from $516,078 to $602,994. This change and its significance are not easy to grasp. In comparison, most people can readily under-

stand the expression, "Revenues rose during the quarter by 17%." Remember the formula for percentage of change: *($602,994 – $516,078) ÷ $516,078 = 17%.*

— Percentages bring the numbers to life when used in comparison form. For example, if revenues rose $516,078 to $602,994, what if costs and expenses rose from $438,540 to $485,916? What does this mean? This can be dramatically contrasted with the statement: "While revenues rose in the quarter by 17%, costs and expenses rose only 11%. This explains the dramatic increase in net operating profit from 15% last quarter to 19% this quarter." (Net is calculated by subtracting costs and expenses from revenues: *$516,078 – $438,540 = $77,538*; and *$602,994 – $485,916 = $117,078*. The calculation of net return is the result of dividing net profit by revenues: *$77,538 ÷ $516,078 = 15%*; and *$117,078 ÷ $602,994 = 19%*.)

2. *Describe change in terms of the "number of times" greater or smaller rather than listing the dollar amounts.* Avoid dollar values in narrative sections whenever possible. Given the previous example, it is a better form of communication to explain that "revenues increased by 1.17 times over the past quarter (*$602,994 ÷ $516,078*)." It is also effective to describe net profit changes as: "Net operating profits surpassed the previous quarter by 1.51 times." (*$117,078 ÷$77,538*).

3. *Express the important features of numerical trends in narrative-based ratio expressions.* Another version of the non-dollar expression of results is the use of ratio-type expressions in place of dollar values or percentages. For example, rather than stating that revenues this quarter were 1.17 times higher, the ratio version is "the change in revenues was 1.17 to 1 over the prior quarter." Although this is the same as a percentage increase, it is easier to digest than percentages. Many readers are going to be uncomfortable with any form of math, including percentages; but they can easily understand and appreciate what "1.17 to 1" means. It is the same thing but expressed in a less intimidating manner.

Combining Narrative and Financial

In any report where you combine narrative and financial information, you face a dilemma. You want to reduce exposure to the numerical information, but you often are required to include and disclose the full range of results. This is where appendix placement is a good idea. Unfortunately, separating the financial data from the discussion is not always effective. This is where two additional reporting tools are effective: Annotation and highlighting.

Annotation is an effective highlighting tool for bringing numbers to life, and for combining narrative and financial sections. For example, if inclusion of financial information is unavoidable, you can use annotation to explain the numbers. Consider the following summarized profit and loss statement:

Revenue	$14,862,300
Cost of goods sold	– 7,992,500
Gross profit	$ 6,869,800
General Expenses	– 4,481,500
Net operating profit	$ 2,388,300
Other income and expense	107,900
Net pretax profit	$ 2,496,200
Less: Liability for taxes	– 743,900
Net after-tax profit	$ 1,752,300

By itself, this summary may be interesting, but it does not reveal the significance of the changes in trends. You can use annotation to make this summary far more interesting. Figure 8.2 shows how this is accomplished.

		19% increase over last
Revenue	$14,862,300	quarter
Cost of goods sold	- 7,992,500	
Gross profit	$ 6,869,800	46.2% gross profit
General Expenses	- 4,481,500	
Net operating profit	$ 2,388,300	
Other income and expense	107,900	16.1% versus only 11.5% last quarter
Net pretax profit	$ 2,496,200	
Less: Liability for taxes	- 743,900	
Net after-tax profit	$ 1,752,300	
		highest net on record

Figure 8.2: Annotated financial data

You can also highlight specific information through many devices, including italics, larger fonts, underlining, and call-out. For example, you may use a call-out to provide details of the cost of goods sold, or to provide the comments shown in annotated form in Figure 8.2. When you must combine narrative and financial information, it can be made interesting.

Too many reports simply list the data. For example, the following paragraph is a typical reporting of financial data:

> Quarterly revenues were $14,862,300, 19% higher than last quarter. Cost of goods sold was $7,992,500 and gross profit was $6,869,800, or 46.2% gross margin. General expenses were $4,481,500, for a net operating profit of $2,388,300, or 16.1% (compared to only 11.5% last quarter). Other income was $107,900 and pretax profit was $2,496,200. After deducting tax liabilities of $743,900, the net after-tax profit was $1,752,300, the highest net profit on record.

This report would be far more interesting in either annotation or highlighting in some form. The problem with a narrative summary is that it provides obvious information (such as the cost of goods sold *and* gross profit) but does not comment on its significance. Consider this revised paragraph:

> This was a quarter of new records. Quarterly revenues 19% higher than last quarter and gross margin remained steady at 46.2%. Internal controls produced an impressive 16.1% net operating profit, 4.6% higher than last quarter. Net after-tax profit was $1,752,300, the highest net profit on record. (Details provided on next page.)

The next page contains the relatively dry numerical summary, preferably with the annotated commentary. This shorter paragraph is far more effective because it explains what is significant in the latest report and does so without listing any of the dollar values. This is acceptable because the financial summary is shown on the following page, along with the emphasis where attention belongs. A comparison between both versions of this paragraph demonstrates how a dull, monotonous report can be livened up and made interesting.

Graphics in Reports

Including graphs in your reports makes a tremendous amount of difference. In the past, before spreadsheet-based graphics were possible, report preparers had to rely on a company's art department or outside graphics companies to prepare charts. This was time-consuming and expensive and therefore, most internal reports lacked graphics or included only crudely drawn approximations. The value of graphics is profound. Comprehension is vastly improved when information is presented graphically rather than in narrative or numerical form:

> Because our eyes detect a limited set of visual characteristics, such as shape or contrast, we easily combine these characteristics and unconsciously perceive them as an image. In contrast to "attentive processing"—the conscious part of perception that allows us to perceive things serially—pre-attentive processing is done in parallel and is much faster. Pre-attentive processing allows the reader to perceive multiple basic visual elements simultaneously.[2]

When an audience is presented with a visual summary of your message, it is not only comprehended more clearly, but remembered more clearly as well. In constructing graphs using the easy facilities provided in word processing or spreadsheet programs, remember a few guidelines:

---

2 Schwabish, J. (2014). An Economist's Guide to Visualizing Data. *The Journal of Economic Perspectives*, 28(1), 209–233.

1. *Make sure the graphic is used only for important information.* Avoid the temptation to graphically illustrate everything, even the most obvious data. Reserve graphic treatment to summarize the numbers, to visually show a trend, or to compare two or more trends.
2. *Select the most appropriate graph.* A single bar chart is applicable to a single trend over time; and a double bar graph is the best to use in showing two or more factors for several groups. Earlier in this chapter, this was used to show sales comparisons for several divisions. This provided not only the change within each division, but a visual comparison between divisions as well. A line graph is useful for tracking a trend over many months or years. Finally, a pie chart shows a circular summary of information; for example, where does the average dollar of revenue go each year? The pie chart breaks down this information into degrees of the circle. This requires two separate calculations: Percent of the total and the breakdown into degrees.

Formula: Percent of the Total

$$V \div T = P$$

where: V = value
T = total
P = percent of the total

On a spreadsheet:
A1 T
copy A1
paste to A2, A3, A4, etc.
B1 V(1)
B2 V(2)
B3 V(3)
B4 V(4)
C1 =Sum(B1/A1)
copy C1
paste to C2, C3, C4, etc.

Example: Total revenue is $14,862,300. This is broken down into the cost of goods sold, general expenses less other income, liability for taxes, and net after-tax profit. Each of these represents a percent of the total:

Cost of goods sold	7,992,500 ÷ 14,862,300 =	53.8%
General expenses and other income	4,373,600 ÷ 14,862,300 =	29.4
Liability for taxes	743,900 ÷ 14,862,300 =	5.0

Net after-tax profit	1,752,300 ÷ 14,862,300 = 11.8
Total	100.0%

Next, this breakdown is applied to calculate degrees of the circle:

Formula: Degrees of a Circle

$$P \times 360 = D$$

where: P = percent of the total
D = degrees

On a spreadsheet:
A1 P
B1 =SUM(A1*360)

Example: You are preparing a pie chart to show how each dollar of revenue is spent during the quarter, based on the latest income statement. Total revenue was $14,862,300. The 'P' values were:

Cost of goods sold	53.8%
General expenses and other income	29.4
Liability for taxes	5.0
Net after-tax profit	11.8
Total	100.0%

Applying the formula, first convert the percentages to decimal form and then multiply:

Cost of goods sold	.538 × 360 = 194
General expenses and other income	.294 × 360 = 106
Liability for taxes	.050 × 360 = 18
Net after-tax profit	.118 × 360 = 42
Total	360 degrees

After you calculate the degrees, those values are converted to the pie chart formatting feature and the pie chart is produced. Figure 8.3 shows the outcome.

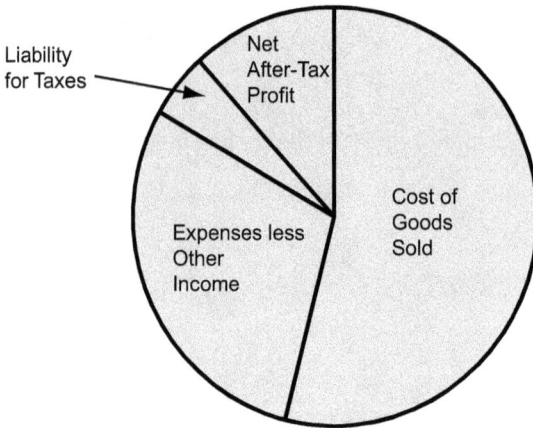

Figure 8.3: Pie chart for financial report

This chart format is particularly dramatic because it shows how each dollar of revenue is applied. While the numbers by themselves and even the trends in the numbers represent meaningful information, nothing compares to a visual summary.

The next chapter focuses on a report that every manager contends with, and that most managers dread: The budget. The chapter demonstrates how you can thoroughly document assumptions to turn a budget burden into an effective management tool.

Chapter 9
Budgeting Calculations: Assumptions and Pro-rations

Budgets have been described as a list of priorities and "political" documents.[1] This is true because, during setting priorities within an organization, the individual who has authority to prepare and approve budgets determines where money is going to be spent, which segments, or departments will be expanded or allowed to hire more employees, and which projects will rise to the top of the list.

As an aspect of management accounting concerned with producing financial information for decisions, budgeting:

> ... is usually an annual, cyclical process (although budgets can be prepared for longer or shorter periods). However, with a sound system of budgeting in place, the other management account-ing activities—whether short term ad hoc investigations or longer-term strategic plans—become much more straightforward and are built on secure foundations ... Budgeting, in fact, is not just a financial exercise, but, as with management accounting in total, an integral part of management.[2]

As a manager, do you prepare a hands-on version of a budget, or rely on others? It makes little sense to delegate the departmental or segment budgeting process and its approval to others. Every manager has a vested interest in setting goals for the coming year, overseeing a budget designed to meet those goals, and establishing internal controls to ensure that expenses do not outpace revenue and exceed appropriately established limits.

However, many budgeting steps can be delegated to subordinates. "Effective del-egation can build confidence in managers, and it shows subordinates that their man-agers trust them ... The underlying assumption of effective delegation is that manag-ers are confident in their employees, themselves, and what they are doing."[3]

Whether you prepare a budget directly or supervise others, or even if your budget is prepared outside of your area of control and presented to you for review, there are a few very key considerations to remember when going through the budgeting process:

1. *Every budget should be based on sensible assumptions.* The assumptions are the underlying justifications for every budget item. It is not effective to simply increase the previous year's budget by five percent without these assumptions. That process—which is the most common way that budgets are prepared—is

---

1 "A budget is a statement of priorities, and there's no more political document."—Edward K. Hamil-ton, *New York Times*, February 9, 1971.
2 Cook, A. (1995). Management Accounting. *BMJ: British Medical Journal, 310*(6976), 381–385.
3 Jin, G., & Merritt, D. (1993). Delegating as a Component of Managing Effectively. *The Journal of Tech-nology Studies, 19*(2), 60–63.

DOI 10.1515/9781547400638-009

entirely ineffective and does not establish any spending controls. It only makes past spending excesses permanent.

2. *It makes no sense to prepare a budget for more than one fiscal year.* The purpose of the budget is to set goals for net profits, based on controlling expense levels. However, because the environment is constantly changing, even a full-year budget cannot be expected to accurately predict the future.

3. *Six-month revisions are essential, but also must be based on reality.* Assumptions need to be updated. A revised budget is prepared too often to absorb mistakes made in the original budget. Rather, a revision should update logical assumptions to keep the budgeting process effective and valuable.

4. *The true purpose of budgeting is to monitor progress.* The budget should serve as a financial expression of future goals, based on underlying assumptions. This means that the budget is the detailed portions of a larger goal. The purpose is not to ensure that actual expenses will fall under budget in each case, but to provide yourself with a means to check progress, recognize internal control problems, and fix those problems as they arise.

5. *Even with proper use of a budget, it remains a political document.* The proper preparation and use of a budget cannot get around the fact that it is a political statement of priorities. Competition within an organization for a larger budget and for more status and power define how budgets are prepared and revised. Ultimately, budgets are also used to identify success and failure within an organization.

Documenting Your Assumptions

When you review a budget, do you just check the variances? These important outcomes, the differences between the "budgeted" and "actual" totals, are only one of the factors you need to keep in mind. An equally crucial piece of information is the underlying assumption. For example, do you assume the expense level will be the same as last year's? Are you factoring in any increases? Are you using the most applicable base?

While variances are important, so are the assumptions. Some typical assumption bases you need to use in preparing a budget (or that you should expect to find when reviewing budgets prepared by someone else):

1. *Relationship to revenue growth.* Some general expenses are going to change based on known or planned changes in volume of sales. For example, if your company is acquiring another company, you are going to increase your floor space, number of employees, and internal expenses (shipping, telephone, office supplies, telephone, utilities). Even though expenses are not "direct" in the same way as costs, the overall level of expenses is going to rise, or fall based on changes in revenue.

2. *Expenses varying by numbers of employees.* One way to reliably budget many kinds of expenses is by the number of employees. For example, you can analyze your office supply expenses by expense per person. To check this, compare several years worth of supply and employee levels. The same rule may apply to printing, delivery, telephone, and several other administrative expenses.

3. *Marketing expenses such as auto and travel.* The sales and marketing expenses often are listed separately from general and administrative expenses because there is a strong correlation between volume of sales and marketing expenses. If your company is initiating a new product line, creating new incentives for salespeople, expanding a supply chain, and other changes in how products or services are delivered, then marketing expenses must be adjusted realistically at the same time.

4. *Inflation and economic factors.* Most often overlooked in the budget are the unavoidable effects of inflation and other economic changes. The price of gasoline affects auto and sales expenses; utility rate increases directly affect your expense totals; and if you deal overseas, foreign currency exchange fluctuations are also going to have an immediate effect on virtually every line of your income statement.

5. *Efficiency from consolidation and internal controls.* If you have put new internal controls in place, you naturally expect to generate savings; this should be reflected in the budget, which is where you monitor the expense. If you have consolidated departments or segments, you would also expect to realize efficiency and lower expenses as a result. These changes must be folded into your budget assumptions.

6. *Changes in internal procedures, including improved efficiency.* If you were previously relying on outside services for expenses such as printing or payroll, and this year you are bringing these services in-house, you will have immediate changes in expense levels. Any internal change in procedures will have an impact, hopefully a reduction in expenses. This should be allowed for in the budget assumptions.

7. In preparing a budget, the detailed expense categories should be fully documented by assumption base, and then broken down by month. To what extent this process is a "best guess" depends on the strength of information. This is where math comes into the picture. Most expense categories can be budgeted on known information, historical trends, or revenue forecasts. However, some expenses simply cannot be based on solid assumptions. This raises the question: If you have no sensible assumption for budgeting an expense, how do you proceed?

The answer is that you must rely on math to create a best guess budget. For example, telephone expenses have been rising over the past year, but you cannot identify a base for building an assumption. Lacking any reliable assumption base, you turn to the *trend* to create your budget. During the past year, this expense was booked in the following amounts for your department:

January	$1,221	July	$1,777
February	1,242	August	1,804
March	1,231	September	1,816
April	1,282	October	1,990
May	1,336	November	1,910
June	1,545	December	1,915

The general trend throughout the year was upward. However, there is no recognizable seasonal or cyclical change to blame, so one conclusion might be that the expense is not being properly controlled internally. The number of employees has remained unchanged and no marketing activity is generated in your department, so changes in revenue and marketing cannot be blamed for this increase. Telephone rates did increase to a degree but cannot account for the significant change experienced by your department. Initial research indicates that another change in billing rates is going to occur during the first quarter estimated at 5%, so this must be built into the budget. However, you also believe that some abuses have occurred during the year, accounting for the growth in rates from an average of about $1,250 per month up to more than $1,900, an increase over 50%.

Addressing the Expense Issues

The budget should serve as a means for monitoring the solutions you put into place and making judgments about whether they were effective. This requires taking some chances. If your internal controls are not effective or you are mistaken about how effectively you can keep expenses low, you will experience unfavorable variances. Some managers will allow budgets to be set too high just to avoid these, whereas others view unfavorable variances as an opportunity to identify and fix problems.

You take the following actions to improve internal procedures:
- Announcement of a new monitoring process accompanied by a reminder that personal use of telephones is a violation of company policy.
- Establishment of a manual phone log to be used for all long-distance use of the phone.
- Improved internal procedures for identifying long distance calls by individual telephone.

These procedures should be manageable because no marketing activity comes from the department. The only long-distance calls should be made to branch offices or to the home office. It is also possible with modern telephone systems that fixed fee telephone usage is possible, but this is considered too expensive for a non-marketing department. Your budgeting assumptions include the following:

- A reasonable base for telephone usage is $1,200 per month at the beginning of the year, and a 5% increase to $1,260 beginning in March.
- The internal controls announced and put into place are expected to have an immediate effect and reduction of monthly charges.
- A quarterly allowance of $150 will be added to the beginning of the first month in the quarter for long distance usage based on documented phone log entries.

To properly document your budget for the year, all these components should be included in the monthly budget:

Month	Base expense	Rate increase	Long distance	Total
January	$1,200		$150	$1,350
February	1,200			1,200
March00	1,200	$60		1,260
April	1,200	60	150	1,410
May	1,200	60		1,260
June	1,200	60		1,260
July	1,200	60	150	1,410
August	1,200	60		1,260
September	1,200	60		1,260
October	1,200	60	150	1,410
November	1,200	60		1,260
December	1,200	60		1,260
Total	$14,400	$600	$600	$15,600

While you cannot expect telephone expenses to conform exactly to this budget, you have the assumption base as a starting point. This means you can monitor each month's actual expenses against the budget and identify where problems continue to occur. If the budget turns out to be inadequate for the legitimate monthly usage, having a well documented assumption base for your budget allows you to identify why this is so, and to fix the budget at a six-month revision.

Without the assumption base, you simply have no way to know whether expenses are too high or not. This budget also allows you to monitor the success of your new internal procedures. If the very high monthly usage rate does suddenly fall back to the base rate of $1,200 per month, that tells you that past abuse of the system has stopped due to the announcement of new monitoring. This set of budget assumptions is defendable because it completely documents how it was arrived at. Another method—taking the past year's total expense and dividing it by 12 to arrive at a monthly budget—is much less effective. When expenses run over the month's budget, how do you explain it? The strength and detail of the assumption base gives you the means to explain variances, and when you lack an assumption base you cannot possibly defend or explain the budget. The development and documentation of assumptions should be one of the budget rules that everyone follows:

> People tend to expect both too much and too little of budget rules. Because they are more art (or perhaps, craft) than science, it is impossible to derive some ideal set of rules through theory. Because they involve many arbitrary elements, they are also easy to disparage. And yet, an orderly decision process requires rules, written or unwritten.[4]

This observation applies to how assumptions are developed. An arbitrary budget cannot identify where controls were ineffective, whereas a thoughtfully developed set of assumptions provides intelligent means for later evaluating variances.

The purpose in budgeting is to create a mathematically detailed and itemized rationale for the monthly expense total. This is used to compare budget to actual and to identify areas where (a) problems exist and require fixing, (b) the budget was inadequate, or (c) the realistic level of expense was not anticipated when the budget was prepared. Any of these outcomes is acceptable because it improves information and gives you what you need to improve in the budgeting process. The only undesirable outcome is when assumptions are not documented well enough to provide an explanation for why the budget falls short.

Prorating Expense Estimates

One process that comes up often in budgeting is proration, the division of a single item among many different departments or segments. The formula for proration was introduced in an earlier chapter. The purpose for raising the topic here is to provide the means for making proration of expenses during the budget process.

Example: Your company is budgeting by department and the accounting department's manager is creating a basis for proration of several expenses. Among these are facilities rent, utilities, and local taxes. Assignment of these expenses to each department is a debatable practice. Some organizations undergo an extensive process of assigning expenses such as these, often arbitrarily. There is no opportunity for a department to control or reduce the expense, so why prorate the budget? It makes more sense to create a corporate budget shell for expenses that cannot be prorated. However, in those cases where the decision to prorate has been made, it is important to develop a logical method. In the case of both rent and utilities, the decision was made to prorate monthly expenses based on square feet occupied by each department.

---

4 Penner, R., & Steuerle, C. (2004). Budget Rules. *National Tax Journal, 57*(3), 547–557.

Some square footage in the building cannot be assigned, including hallways, meeting rooms, and utility areas. These are ignored; only the square footage of each department is added up, and a total developed for this purpose. Last year, rent was $4,400 per month and utilities averaged $750. The square footage was calculated as:

Department A	1,100
Department B	950
Department C	2,235
Department D	1,570
Total	5,855

These totals are assigned a percentage of the total:

Department A	1,100 ÷ 5,855 =	19%
Department B	950 ÷ 5,855 =	16
Department C	2,235 ÷ 5,855 =	38
Department D	1,570 ÷ 5,855 =	27
Total		100%

These prorated levels are to be assigned to rent and utilities each month. The decision to alter the monthly levels should depend on historical changes. For example, rents may be increased annually based on changes in local property tax assessments or upon expiration and renewal dates of the lease. Utility costs should be carefully analyzed to approximate seasonal increases or decreases based on weather.

Using proration for companywide expenses is problematical as a budgeting method. A basic rule worth observing is that a department should have the full capability to control every item in its budget. Failing this, the budget remains a strictly political or bureaucratic tool with no lasting value. For example, a manager might be criticized because an administrative department is "costing too much" to the company. However, if the annual cost includes prorated expenses in the budget, this is an unrealistic and disingenuous criticism. It might be fair if the department's budget included only those items the manager was able to control; but if prorated expenses are placed into the budget by way of an annual journal entry, it only causes problems and reduces the accuracy of financial review.

If the purpose of budgeting is to provide managers with the information they need to monitor internal controls and to hold down expenses, it is a worthwhile pursuit. In many organizations, the "information supply chain" is corrupted because budgets are based not on a manager's realistic capability to reduce expenses, but on a series of assigned budget items that cannot be controlled at all.

Calculating Variances

The ultimate purpose of the budget is not to avoid variances that must then be explained. It is to identify problem areas as early as possible so that corrective action can be taken. The budget is a mechanism for expense control, and not simply a compliance device to identify expense levels in advance and then adhere to them:

> The budget provides management with a method for evaluating the performance of managers. The assumption is that since the budget represents the desired state and the manager is aware of the budget, he should make those decisions necessary to minimize deviations from the budget over time.[5]

How is this possible without realistically set assumptions? In studying and explaining variances, the assumptions must be based on the well understood purpose of the budget itself.

Calculating budget variances is the key to monitoring expense levels and controlling them effectively. However, the budget is useless without the required follow-up action. A monthly report summarizing actual, budget and variances must be prepared; and then you need to determine what you need to do to fix any discovered problems.

To calculate a budget variance, first list the expenses as they have been reported for the month; and then list the month's budget items. This raises an important point: Do you study only the month's numbers, or do you base your budget analysis on year-to-date?

The advantage of the monthly analysis is that it studies progress for that month. However, it ignores past variances and does not allow you to explain that some items are merely timing differences. An actual expense might not be booked in the exact month it is budgeted. For this reason, an adjusted year-to-date analysis of both actual and budget makes the most sense. To calculate year-to-date expenses, add the current month's total to the prior year-to-date total.

Formula: Year-to-date Expense

$$C + Y = E$$

> where: C = current-month expense
> Y = prior year-to-date expense
> E = year-to-date expense

---

5 Searfoss, D., & Monczka, R. (1973). Perceived Participation in the Budget Process and Motivation to Achieve the Budget. *The Academy of Management Journal, 16*(4), 541–554.

On a spreadsheet, list each expense item in its own column and enter:

A1   C
B1   Y
C1   =SUM(A1+B1)

Example: Your prior year-to-date expense for telephone was $2,815. This month's expense is $1,192. The year-to-date expense is:

$$\$1,192 + \$2,815 = \$4,007$$

To calculate year-to-date budget, add the current month to the prior year-to-date total.

Formula: Year-to-date Budget

$$C + Y = B$$

where: C = current-month budget
Y = prior year-to-date budget
E = year-to-date budget

On a spreadsheet, list each budget item in its own column and enter:

A1   C
B1   Y
C1   =SUM(A1+B1)

Example: Your prior year-to-date budget for telephone was $2,550. This month's budget is $1,260. The year-to-date budget is:

$$\$1,260 + \$2,550 = \$3,810$$

The variance report consists of identifying significant variances and explaining them, whether favorable or unfavorable. You need to identify a "significant" variance as a starting point. You have no need to explain minor variations, which are going to occur in just about every expense category. The definition relies on the overall size of your expenses. For example, if your monthly expenses average $3,000, a variance of $100 might be considered significant. If your monthly expenses average $30,000, a $100 variance might not be worth analyzing. A typical definition of a "significant variance" is:

Any variance of $300 or more and 5% or more above or below the budget.

This definition requires two tests. If the variance is above $300 but under the 5% threshold, it should not be analyzed under this definition. To calculate a variance, compute the difference between actual expenses and budgeted expenses year-to-date.

Formula: Expense Variance

$$B - E = V$$

> where: B = year-to-date budget
> E = year-to-date expense
> V = variance

On a spreadsheet, enter for each expense line item the following:
A1   B
B1   E
C1   =SUM(A1-B1)

If the value of 'V' is positive, then it is a favorable variance, meaning that the budget total is higher than actual expenses. Even a favorable variance should be studied and explained if it is significant. It could be caused by a timing problem between months, or by faulty assumptions during budget preparation.

   If the value of 'V' is negative, it is an unfavorable variance and must be explained, if it meets the definition of a "significant variance."

Example: You have defined a "significant variance" as any variance of 5% or more higher or lower than the year-to-date budget, and of $100 or more. At the end of March, your year-to-date budget for telephone expenses is $3,810, and actual telephone expenses year-to-date are $4,007. Your budget variance year-to-date is:

$$\$3,810 - \$4,007 = -\$197$$

To calculate the percent by which the variance is higher or lower than the budget, divide the year-to-date variance by the year-to-date budget.

Formula: Percent of Expense Variance

$$V \div B = P$$

> where: V = year-to-date variance (favorable or unfavorable)
> B = year-to-date budget
> P = percent of expense variance

On a spreadsheet:

    A1 V
    B1 B
    C1 =SUM(A1/B1)

In the example, the outcome is a "significant variance because it meets both tests. It is higher than 5% of the year-to-date budget and the amount is greater than $100. It requires an explanation:

$$\$\text{-}197 \div \$3,810 = -5.2\%$$

Revising Budgets

The budget is not a process you execute once or twice per year and then forget. Used effectively, the budget is a monitoring tool for controlling expenses and for spotting trends early enough to curtail excessive losses. Too often, budgets are imposed on managers with an expectation that somehow, they will keep expenses within the mandated constraints, but without also providing the control needed to accomplish that goal. This is a mistake. To work as intended, the budget must be properly documented, used every month, and upon discovery of an expense overrun, action must be taken.

No matter how much ground work goes into the budget, you are going to discover errors and changed circumstances. It is impossible to anticipate the future for more than a few months due to ever-changing competitive, market and economic conditions. While an annual budget is prepared at the beginning of each year, a six-month revision is necessary. Remember, the purpose of the budget is to set a standard for keeping expenses under control. After six months, you will spot new trends and recognize flaws in the original budget. The reason for a revision is to reset the budget to a more accurate base, but not to move expense levels up to accommodate out-of-control expense spending.

Expenses should be revised when:
1. *Conditions in your company have changed, making the original budget obsolete.* The market and competitive factors you face continually are constantly changing, meaning that your company is changing as well. In a free enterprise environment, change is a constant and even in a controlled system, nothing remains the same for long. Even governmental management involves continual adjustment of budgets as revenue moves upward or downward due to political and economic conditions. In other words, no company can operate for long in isolation. A 12-month budget is essential for setting annual goals, but after six months it is almost always outdated.

Any expense whose original budget is no longer applicable to circumstances needs revision. For example, if revenues have risen far above expectations and internal staffing has been increased as a result, all expenses budgeted on a per-employee usage basis are now outdated.

2. *Mergers or disposals of operating segments have changed the landscape.* Your company may close a merger or acquisition deal with another, resulting in consolidation of departments in a newly centralized location; or an operating segment may be sold off during the year. Both changes make the existing budget obsolete. Even if the change does not affect your immediate budget directly, newly structured profit and loss assumptions, increased or reduced goals, different levels of proration, and even changes in accounting methods, all mean the current budget must be revised and replaced.

3. *New internal controls or processes have been put in place.* You may experience a complete replacement of internal controls or processes. This occurs due to advice from independent auditors and systems consultants, regulatory or legal mandates, improved internal processing methods, adjustments in the supply chain, and even changes instituted by suppliers or vendors. Nothing is permanent, and managers must continually seek ways to hold down expenses. The budget is the monitoring device, but whenever change is imposed on the operational model, the budget must be changed to reflect the new reality.

4. *Departments have been merged or otherwise changed.* In the interest of efficiency, management may change the definition of internal departments or segments. Simply combining two departments into one invariably results in reduction of the workforce, changes in processing, and an entirely different set of budget assumptions for most line items. For example, if two marketing branches are consolidated, it can reduce the need for separate processing within the administration of the company, reduction or reassignment of the sales force, and fewer personnel for tracking, accounting, and processing departments. Expanding the branch system may create the need for larger departmental payroll and other expenses, and even segmenting of current support staff into new configurations.

Changes such as these make the current budget useless. Every change in the configuration of the company is going to have a direct effect on the budgets of most, if not all departments. For this reason, six-month revisions—at the least— must be assumed as necessary.

The revision should never be undertaken simply to adjust for discovered errors or to absorb unfavorable variances. It is only effective to find ways to fix expense overruns or to replace existing budgets with more realistic assumptions.

The Nature of Revenue Forecasts

The process of forecasting revenue is quite different than that for budgeting expenses. In an expense budget, assumptions are defined by the nature of the expense and only a few bases are applicable. Assumptions using square feet are used to allocate or prorate utilities or printing. The number of employees often determines the budget for office supplies, telephone, and other varying expenses. Items like rent are often predetermined by contract. Another area of assumption base is proration, with some expenses assigned to departments based on square footage or number of employees.

When it comes to forecasting revenue, an entirely different series of assumptions apply, and these are based on the nature of the organization, the product or service marketed, and the kind of competitive environment. Some examples of appropriate assumptions:

1. *Salespersons' activity and history.* In any company relying on a sales force or branch office system, notably those based on commission compensation, many marketing activities promote greater volume. The use of commission incentives or awards can produce exceptional volume in target periods.
2. *Market share and trend.* How does your organization compare to other companies marketing the same products or services? There is a realistic limitation on how much market share a company will be able to capture. Is the market itself expanding or fixed? If sales are based on changes in population within a market region, then this should be taken into consideration when forecasting revenue growth. Is the existing trend positive? If so, will it continue and to what extent? In tracking a revenue trend, remember two key elements: First, market share is finite. Second, the line of a trend is going to begin leveling out in the future; no existing rate of growth can continue indefinitely.
3. *Product or service line expansion.* Are the products or services you offer expanding or remaining at present levels? Without this kind of expansion, revenue growth is going to top out at some point. Even an expanded product or service line also requires time to show up on the books. Expanding the market may consist of gaining more customers or offering customers more in the future than in the past.
4. *Seasonal variation.* In preparing your forecast, remember to take seasonal variation into account. Most organizations experience high and low volume periods, especially those marketing products. You cannot realistically expect to spread a forecast evenly throughout the year. It makes more sense to study past volume by month and then apply the curve of the seasonal change to your new forecast as well. This helps not only to place the right degree of forecast into the right season, but also to compare year-to-year outcomes to better track and anticipate revenue growth.
5. *Planned acquisitions or sales.* If your company is going to acquire or merge with another company during the year, the forecast may need to be prepared on a pro-

visional basis. One outcome is expected to occur if the merger goes ahead, and another is expected if it does not. The same applies if you expect to sell off a segment during the coming year. If the sale does not take place, revenue should be forecast with existing assumptions. If the sale does occur, revenue (as well as costs and expenses) would be expected to decline immediately after the sale occurs.

6. *Existing trend.* The existing trend dictates all the assumptions going forward. The trend in revenue cannot be based on the assumption that the existing trend is going to continue without slow-down, so the trend must be realistically leveled out. This is also modified by the addition of new markets, products, or services, and changes in competition (due to mergers or sales).

When you prepare a revenue forecast, it also requires tracking of direct costs. These are tied directly to revenue and accordingly, should be easy to estimate. However, some factors are going to alter your direct cost estimates, including:

1. *Changes in the long-term trend.* Have your direct costs remained constant for the past few years? Or have they been changing? For many internal and external reasons, the gross profit you book may be evolving over time. The trend should not be assumed to remain unchanged unless you have strong proof that it is a constant. In organizations providing services rather than products, direct costs are less of a factor than for those manufacturing, transporting, and selling tangible products.

2. *Known changes in merchandise costs.* Are your suppliers continuing to provide merchandise to you at the same cost level as in the past? Have cost increases been reflected in higher retail prices? Have your manufacturing or labor costs changed? These matters need to be studied in the cost assumptions. If you are booking a lower gross profit than in the past, should you be marking up your prices to make up the difference?

3. *Changes in the product mix.* If you are selling a more diverse range of products today than in the past, the long-term trend could be unreliable. If newly added products require a different level of costs, and if these new products represent a growing percentage of total revenue, then historical cost ranges are not accurate.

4. *Adjustments in the supply chain and inventory practices.* With global supply chains becoming more common than ever before, you need to consider within the cost of goods sold a broad range of costs and possible changes. These include overseas manufacturing trends, labor issues, warehousing, transportation, and several related supply chain risks that can affect your ability to meet demand and to keep costs under control. It makes sense in an income/cost forecast to include a contingent cost for supply chain risks and possible losses.

5. Many of the forecasting and budgeting tasks require the study of trends, use of averages, and making allowances for change. All of these, like so many of your tasks, require a basic understanding of statistics. The next chapter explains the basic statistical formulas and applications every manager needs to know.

Chapter 10
Statistics for Effective Reporting

Beyond the realm of the familiar math involving financial reporting, marketing, and budgeting, the more complex world of *statistics* must be addressed. A few important management applications require basic statistical skills. This chapter explains these basics and how they apply to you in your role as manager.

In statistics, you deal with a *field* of numbers. This is a grouping of values related in some manner (for example, sales results by division, expense totals during a quarter, or the number of salespeople and their sales volume). These numerical values are arranged in a field so they can be studied and interpreted to create a significant result. In presenting a field of numbers in a report, they have no meaning. But if these are manipulated through applications (such as mean, median or mode, which display the importance of the outcomes), then those numbers are interpreted and can be explained with meaningful summaries.

Statistics is based on how numbers, set up in fields, are manipulated so the results can be interpreted. You might say that the numbers by themselves represent passive results; the statistical interpretation becomes the active result.

Some observers believe that business statistics are relatively simple, but that is not the case:

> Official statistics traditionally make a distinction between statistics collected for business, also called "business statistics," and statistics collected from households. This separation is clearly reflected [in] publications on survey methodology. But is it a fundamental distinction? Are business statistics a special case? ... Everything is intricate, and this complexity is fundamentally due to the fact that the entity observed—the business—cannot be portrayed simply.[1]

As a business manager, your goal should be to find ways to simplify and clarify information while improving accuracy, even though many variables and complexities present challenges. Statistics are too often used for the opposite purpose, alienating the recipient rather than providing an enlightened and improved summary of facts. The statistical processes you use should accomplish the goal of improving information instead of complicating it.

---

1 Rivière, P. (2002). What Makes Business Statistics Special? *International Statistical Review/Revue Internationale De Statistique, 70*(1), 145–159.

DOI 10.1515/9781547400638-010

Management Application of Statistics

Most managers are continually awash in numbers, projections, trends, and reports. You cannot avoid statistics, but you do need to master them. This does not mean you need to get an advanced degree, but it does require knowledge about a few important formulas. You are going to need statistics in budgeting, marketing testing, risk management and virtually any other facet of management involved with judgment about dollar value, employees, trends, and profitability. By definition, *statistics* is a set of information based on numbers. As a process, statistics is the method of analysis. As a set of numerical values, statistics are the conclusions you get from manipulating those values, hopefully with an eye to fair results and conclusions:

> Reference to "reality" is a commonplace among both producers and users of statistics. This "reality" is understood to be self-evident: statistics must "reflect reality" or "approximate reality as closely as possible."[2]

The process of defining reality of its approximation is necessary to test a sample of a larger base, because it is not always practical to study all your data. For example, if a product is tested in 15 markets and involves over 10,000 potential customers, you are going to generate a lot of information during a testing phase. With a properly designed survey, you can test a sample of the larger population. The proper use of a statistical sample requires that you identify methods to accurately test a small sample and ensure that it represents the overall population accurately.

Example: A food company wants to test a new product before mass producing it. A test run is set up in several retail outlets. All the tests are performed at 5:30 p.m. on weekdays. Customers are asked to try a sample and answer three short questions:
  6. Do you like the taste?
  7. Would you buy this brand instead of the brand you now use?
  8. What is more important to you, taste, price or ingredients?

The results of this survey indicated that the product would do very well. However, when the product was placed on a broader market, results were dismal. The new product failed to compete with other brands and turned out to be an expensive mistake.

The mistake was in the formulation of the test itself. At 5:30 p.m. on weekdays, most of the shoppers were spouses on their way home from work who had been given

---

2 Desosières, A. (2001). How Real Are Statistics? Four Possible Attitudes. *Social Research, 68*(2), 339–355.

a short list of necessities to buy. They were not the family's primary shopper. The test sample consisted of shoppers most attracted to taste, whereas the primary shopper was more concerned with healthy ingredients and had a more developed brand loyalty. On that basis, the new product did not perform well in the larger market.

This example demonstrates that picking an accurate sample from a larger population is a very difficult task and demands great care. A related problem is how questions are structured. The selection of words and phrasing often affects how a person will answer. This is a continual problem in political surveys. Consider the following two examples of how the same question can be asked to affect the outcome:

Version 1: Are you in favor of improved road systems in the county?

Version 2: Do you think your taxes should be raised 30% to pay for new works projects?

Both may be part of a survey aimed at the same sample and concerning the same political issue. Version 1 is going to get a much higher rate of 'yes' answers than version 2, because emphasis is on road improvement. In version 2, emphasis is placed on tax increases.

These issues point out some of the many problems in the use of statistics. They can be used to alter the outcome of a data test, so that whether you are working with market surveys or the result of financial data, finding an honest and consistent method for analyzing data is not as easy as many people think.

Your management task involving statistics includes three keys:

1. *How and where do you gather your data?* Some information is obvious, such as financial results or manufacturing productivity. Other information is more elusive, such as customer preferences or the forecasting of future production. If you need to know the size of a department or an average monthly expense, the information is easily obtained. If you want to know something in the future, you can use statistics to make estimates, but you are dealing with the unknown. The "facts" are vastly different.

2. *How do you use or arrange facts?* The example of the product marketing survey makes the point that focusing on the wrong customer is going to affect the outcome. That test would have been more reliable if it had been performed at many different hours and days. Just because a sample produces a specific result does not mean the entire population will duplicate it. Accuracy relies on dedicated work in identifying the right facts and how and when they are produced.

3. *What is the right question?* The food example focused on taste but did not ask the right question. Perhaps a more reliable series of questions would have been:

    a. Are you the primary food buyer for your family?

    b. If so, what is the most important factor affecting what you buy? (Choices: taste, price, ingredients, packaging, product/brand loyalty.)

c. How does this product compare? Would you buy it instead of your current brand?

The accuracy of a sample is the most difficult aspect of a survey, because if you do not have the right sample then the results are meaningless. Second, what you test and how questions are posed will also affect the outcome. It is very difficult to come up with a neutral question, which is why questions are better designed when they eliminate the sample. For example, asking a grocery store buyer if they are the primary buyer would eliminate those who are not, making the results more reliable. In the example of the road improvement tax issue, a first question might have been, "Do you plan to vote next Tuesday?"

It also is going to affect results when you pick one location over another. Is your political sample taken at a local university campus or in a retirement community? Obviously, these two populations are vastly different, pointing to the need to sub-divide a sample by age group, income, and political affiliation. This is not unique to political questions; the same distinctions are needed in all types of statistical sampling tests.

Statistical sampling is based on the rules of probability; the statistical inference is a testing of a sample to ensure that the sample is representative and that the outcome's probabilities are accurate. If the sample and test are well defined and accurately developed, you can estimate the probability of outcomes. The "law of large numbers" is used in many applications. The 50–50 choice is the best example of how this works. If you flip a coin a few times, you might get more heads or tails; but if the flip the coin thousands of times, the average outcome will become increasingly close to the expected 50–50 result.

This rule applies to more complex calculations than the 50–50 of a coin toss. For example, in the analysis of the risk of a loss, a manager needs to decide which risks to insure and which ones to transfer, mitigate or self-insure. The decision should be made on the likelihood of outcomes. A plant accident needs to be insured because over time, there is a high likelihood of occurrence. The threat of piracy or hijacking is relatively low, especially if there is only limited transport of finished goods involved and that is very local.

To ensure that you use statistics in the best possible way, while keeping it simple and as management, you need to identify and impose a few smart ground rules. These include:

1. *Be sure you understand the meaning of statistical terms.* Words like "sample" and "population" have very specific meaning. There are many more terms, and using statistics requires careful distinctions between similar concepts.
2. *Test concepts and outcomes at every stage.* The entire field of statistics demands an exact discipline. Nothing can be assumed, and no outcome anticipated. It is never possible to remain completely neutral, but the scientific method required

for honest statistical analysis also requires that the procedure be followed without seeking a desired result.

3. *Question any outcome that seems wrong.* Your instincts are usually right, and if a statistical result is so far off that it raises questions, you probably need to review your information again and look for errors.

Statistical Averages

In previous chapters, the concept of the simple and weighted average was introduced and explained. In statistics, the use of central tendency is common. However, it is more often called the *mean* which is the same concept. There is a distinction between three statistical ideas: mean, median, and mode. Each of these involves finding various interpretations of mid-points in a field of values, and each is calculated in a different manner.

A distinction has to be made between *numbers* and *values*. A number is a quantity. For example, 4 and 6 add up to the number 10. A value is an entry in a field. For example, a field consisting of 1, 2, 3, 4 and 5 contains five values.

Managers must exercise caution in calculating averages by different meanings; the three methods do not yield equal or reliable outcomes in every case:

> In order to illustrate the use of some of the more common measures of central tendency in the field of statistics, it is usually necessary to refer to frequency distributions based upon different kinds of observations and measurements. This is natural and appropriate, since such measures as the arithmetic mean, the mode, and the median, do not always give equally good representative values for a given set of numerical data.[3]

The *median* is different from the mean (average). It represents the exact mid-point in a group of values (the "field"). You often hear this used in economic reports. For example, you will hear about the average wage, but the median home price.

For example, in a group of seven values, the median is the fourth value because there are three values above and three values below. If the total values are even-numbered, the median is the average of the two values closest to the middle. For example, in a field of six values, the median is the average between values three and four.

Example: If the field includes an odd number of values, the median is the value in the middle.

---

3 Hallerberg, A. (1952). An Illustration of the Common Statistical Averages. *The Mathematics Teacher, 45*(8), 578–581.

If the field includes an even number, the same formula works—sort of. You need to find the average of the two values closest to the middle. Thus, if there are eight values, you need to sort them first: in the field, median is halfway between values 4 and 5.

For example, if you have a field that includes 8 values,

$$2, 3, 5, 8, 15, 20, 21, 24$$

the mean is calculated as follows:

$$(2 + 3 + 5 + 8 + 15 + 20 + 21 + 24) \div 2 = 12.25$$

The median is calculated by:
1. Determining the number of values, F in the field. In our example F = 8.
2. Then you divide by two to find the middle two values:
   F/2 and F/2 +1 represent the values that we wish to average.
   In our case,

$$8/2 = 4 \text{ and } 8 \div 2 + 1 = 5$$

3. So, we average the fourth and fifth values, which are 8 and 15:

$$(8 + 15) \div 2 = 23 \div 2 = 12.5$$

Therefore, 12.5 is the median in our example. The mean is 12.25 and so the median and the mean are quite close, but that is not necessarily going to be the case depending on the nature of values in your field.

If you have a field that includes 2,3,5,5,6,8, then the median is 5 because the two middle values are 5.

In a spreadsheet, if you have the data in a column, the calculation of the median is simple. For instance, for seven values you would use the following to find the median:

A1:A7   F
B1   =MEDIAN(A1:A7)

The *mode* is another variation by which statisticians represent a large field of values. It is the number that appears most often. For example, a field of values is:

$$2, 3, 5, 5, 6, 8, 15$$

In this example, the mode is 5 because it appears twice, whereas all the other values appear only once. If a field contains more than one mode, the answer is the average of the two. For example:

2, 3, 5, 5, 6, 6, 15

In this field, both 5 and 6 appear twice, tying for the mode. In this case, mode is 5.5. The *mean* is identical to the average.

Calculations of median and mode are used for subsequent statistical computations, especially in the study of how values change and are dispersed through a range of possible outcomes. For example, in a marketing study of customer preferences, you expect to receive a range of responses from one extreme to another. In this process, you need to identify the most common response to determine what "typical" preferences are going to be.

Dispersion, Deviation and Variance

In statistics, analysis of *dispersion* is a key concept. Dispersion (also called *spread*) is the degree of difference between the values in a field, and the average of the field. It is of such importance because it identifies how future changes can be predicted. The greater the dispersion you find, the greater the possibility of variance in the future away from the average. Measurement of dispersion is a scientific way to measure predictability.

For example, dispersion in returns from the stock market may reveal significant dispersion between benchmark indices:

> For example, in the calendar year 1999, the NYSE composite return was 9 percent whereas the Nasdaq was up an astounding 86 percent—a spread of 77 percentage points (pps) between these two major U.S. exchanges.[4]

This disparity between exchanges was a symptom of the infamous dot-com bubble lasting from 1997 to 2001. In this period, IPOs in internet-based firms led to widespread speculation and eventually, to a crash: "... the internet was emerging from the academic realm to enter the world stage, only to crash after the dotcom bubble burst following the flop of the commercial (.com) rush to exploit the web ..."[5]

Measuring disparity (whether in stock markets or business performance) may point to emerging problem areas based on current volatility; and one of the best ways to reduce reported volatility in a field of values is by applying a basic statistical standard: removal of the extremes. By taking out the exceptionally high value and the

---

4 de Silva, H.; Sapra, S.; & Thorley, S. (2001). Return Dispersion and Active Management. *Financial Analysts Journal, 57*(5), 29–42.

5 Graham, M. & Dutton, W. H. (2014). *Society and the Internet: How Networks of Information and Communication are Changing Our Lives.* Oxford UK, p. 4.

exceptionally low value, you narrow down the range and make the dispersion easier to identify as well.

Example: You are estimating the number of employees you expect to have in your division for the coming year. Over the past six months, the number ranged between 6 and 37:

Month	Employees
July	6
August	35
September	26
October	27
November	37
December	35

Of these six months, the first (July) is not typical. This may have resulted from a start-up when only a few employees were on hand, thus making it untypical of what you expect for the future. As a result, you exclude this from your total and calculate only the average of the remaining five months.

Using only the five typical-range values reduces future unpredictability, especially in calculating what is called the mean absolute deviation. This measures the degree of distance from each value from the overall mean (average). The first step is to measure the difference itself, and then squaring that difference. The values are:

Month	Employees	Mean	Difference	Square
August	35	32	3	9
September	26	32	6	36
October	27	32	5	25
November	37	32	5	25
December	35	32	3	9
Total				104

The difference is the key element; it does not matter whether it is positive or negative. The next step is to find the mean (average) of the squared values:

$$(9 + 36 + 25 + 25 + 25) \div 5 = 21$$

The result of the above averaging is called the dispersion factor. By itself, the value is meaningless. But it is the first step in calculation of the variance from the mean, and that is used to predict the volatility of future forecasts. The formula for the mean absolute deviation:

Formula: Mean Absolute Deviation

$$((V1 - A)^2 + (V2 - A)^2 + ... (Vn - A)^2) \div n = D$$

where: V_1= first value in the field
A = average of the field
V_2 = second value in the field
V_n = last value in the field of 'n'
n = number of values in the field
D = mean absolute deviation

On a spreadsheet, the values for each cell are:

A1 through A_n	V (one value per cell through cell 'n')
B1	A
	copy B1 and paste for all 'B' cells
C1 through Cn	=SUM(A1-B1)*(A1-B1)
	copy C1 and paste for all 'C' cells
D1	= SUM(C1:C_n)/5

Applying the example, the resulting cell contents are:

	A	B	C	D
1	35	32	9	21
2	26	32	36	
3	27	32	25	
4	37	32	25	
5	35	32	9	

An alternative method for arriving at the same result is the calculation of variance. In this process, you find the square of each value in the field and then divide those results to find the average. Then the square of the original average is subtracted. In the previous example, the average was 32, so the square was 1,024 (32 × 32). The formula for variance is:

Formula: Variance

$$((V1)^2 + (V2)^2 + ... (Vn)^2) \div (n\text{-}1) - (V1 + V2 + ... Vn)^2 = VR$$

where: V1 = first value in the first
V_2 = second value in the field
V_n = last value in the field of 'n' values

N = number of values in the field

VR = variance

On a spreadsheet, enter the following values:

A1 through A_n	values
B1	=SUM(A1:An)/A
C1	=SUM(A1-B1)*(A1-B1)
	copy C1 and paste to remaining 'C' cells
D1	=SUM(C1:C_n)/(n-1)

Variance is an easier and faster calculation, but it ends up with the same answer as the calculation of mean absolute deviation. By whichever route you arrive at this answer, it is an important quantity because it is used to calculate *standard deviation*, which is the ultimate degree of uncertainty about the future values. When used in budgeting and forecasting, this is a method for identifying the degree of certainty in the estimates you use.

Basic Math Review: Square root is the inverse value of the square, or any number multiplied by itself and expressed as n^2. The square root is used in many statistical and geometric calculations and when it is used, the symbol $\sqrt{\ }$ is the proper indicator. This symbol asks the question, "What number, multiplied by itself, equals the principal value?" For example, $\sqrt{25}=5$ because $5\times5=25$.

Statisticians use the Greek lower-case symbol *sigma*, or σ to represent standard deviation. The calculation requires you to first figure the square root of the variance, and then divide by the average of the field. In the previous example, the variance was 26, so deviation is the square root of that value:

Formula: Standard Deviation

$$\sqrt{v}$$

On a spreadsheet:

A1 SQRT(v)

Applying the previous example:

$$\sqrt{26} = 5.1$$

Relying on established formulas in many spreadsheet programs, this is easy to calculate. Excel, for example, uses the preset formula =SQRT(v). Enter 26 within the parentheses:

A1   =SQRT(26)

The standard deviation is next divided by the average of the previous field, to arrive at the percentage known as the dispersion factor. The formula:

Formula: Dispersion Factor

$$\sqrt{v} \div ((V1 + V2 + \dots Vn)) \div n = D$$

where: $\sqrt{v}$ = variance
V_1 = first value in a field of values
V_2 = second value in a field of values
n = number of values in the field
D = Dispersion factor

On a spreadsheet:

A1	=SQRT(v)
B1 through Bn	values
C1	=SUM(B1:B$_n$)/ A
D1	=SUM(A1/C1)

Referring to the previous example, the formula is:

$$\sqrt{26} \div ((35 + 26 + 27 + 37 + 35)) \, 5 = 15.9\%$$

With this percentage, developed from manipulating the values in the field, you can draw meaningful conclusions:

1. The range used was between 26 and 37, excluding an exceptionally low month as non-typical.
2. The dispersion of this range is a value of 11, or the net difference between the highest and lowest number.
3. The variance factor is 26 as calculated by either mean absolute deviation or variance.
4. The calculation produces a dispersion of 5.1 employees (square root of 26).
5. The dispersion in the list is equal to 15.9%.

The significance of this depends on the size of the numbers and the reliability of estimates within the budget or forecast. In calculating the number of employees you expect to hire within the next year, the relative volatility is 14.4% based on recent historical data. In comparing volatility between different account classifications or even among departments, you may find higher or lower volatility, indicating the degree of reliability overall. For those areas with higher volatility, the indication may be that you need to perform a more detailed analysis to bring down the level of volatility.

The example used is a limited one. However, when applied to larger fields of values or to less predictable areas of forecasting, the development of the dispersion factor is very valuable, especially in comparative form. Therefore, an important part of a statistical process is to test for errors in estimates.

Accuracy in Statistics

Any estimate of the future outcome of anything (forecasts, budgets, cash flow, product sales) is going to contain a degree of deviation due to errors in assumptions and estimates. Statistics provides methods for explaining errors as reasonable or within a range of expectations. Measuring the range of deviations quantifies the error so that it can be compared to errors in other processes, or against an accepted standard.

Beyond the few statistical formulas presented in this chapter, many more advanced processes are used to delve into expected levels of error versus actual outcomes. Unfortunately, the more complex the calculation, the less it is going to be understood by management. Most people do not appreciate statistical summaries because they are known to be based on probability analysis and sample testing. In other words, they cannot be used reliably to pinpoint an outcome. Even though no other process does better in narrowing down variation in outcome, statistical analysis is not widely appreciated.

The budgeting process is an excellent example of a struggle managers are engaged in continually. No one can reliably identify the coming year's revenue, expenses or profits. There are simply too many variables in play. Even the budgeting process itself affects outcomes. For example, if you forecast revenue in an especially aggressive set of assumptions, these may filter out to the marketing force in the field and affect their behavior. Some aspects of forecasting and budgeting contain a degree of self-fulfilling prophecy, further complicating how assumptions are developed and applied. The same limitations apply to product testing, customer surveys, cash flow controls, production levels and any other financial or numerical analysis.

Estimates cannot be avoided; they provide management with a means of setting goals and then monitoring progress. Whether the test is for revenue or profits, controls over cash flow, or success in a new product's market, estimates rely on detailed and logical assumptions, follow-up analysis and action, and an appropriate use of statistics. Even if you limit your analysis to relatively simple statistical tools like exponential moving averages or calculation of dispersion factors, you still rely on accuracy in the underlying assumptions. Estimating and related use of statistics provides many benefits to you as a manager. These include:

1. *Risk analysis and identification.* Understanding the range of risks you face and how to deal with them is a pervasive and challenging management problem. With the advancement of global supply chains and reliance on international vendors, manufacturing, storage and transportation, the risk universe has expanded in recent decades. You need to decide which risks to mitigate, insure, transfer or accept without action. The decision relies on the likelihood of loss, potential cost of an event, and your ability to affect the environment. These are big questions and you cannot decide how to address risks (including possible unknown or unidentified risks) without detailed study.

    This requires many forms of estimation, including statistical analysis. Insurance companies use statistical analysis to determine how much to charge in premiums and require in deductible and co-payment levels for a covered risk, and which risks to not even cover. The same approach must be used for risks you decide to mitigate, transfer or accept. You mitigate a risk by changing the environment to reduce exposure. You transfer a risk by passing it on to another entity (a subcontractor, vendor or even your customer). And you accept a risk by determining that a loss is unlikely and that the other choices are cost-prohibitive. Advanced statistical analysis is likely to play an important role in this analysis, and you probably will rely on consultation with a risk management company or consultant. It is instructive and necessary to also understand how statistical analysis is performed to reach a conclusion. Just as your presentations or requests to management require explanation, when you work with someone else, you also need to understand how their conclusions were developed. Risk analysis is probably one of the most important undertakings for managers.

2. *Quantifying the feasibility of a proposal.* Managers are idea people. You are likely to put for the proposals for changes in processes, cost-cutting, internal controls, customer service, and improvements to your supply chain. All these proposals vary in their feasibility, meaning they have practical restrictions, costs, and time elements. These determine whether a proposal makes sense or not. If you approach management with a proposal (or someone reporting to you comes forward with a proposal), these important considerations are present whether expressed or not.

    A common response to an idea is "What is this going to cost?" This is a central question, of course, and the answer is most likely to decide whether an idea is approved or rejected. However, additional questions should be addressed as well. These are "What restrictions or limits apply?" and "How long will it take to implement?" If you keep these questions in mind, it is more likely that you will be able to quantify the feasibility of a proposal. For example, if a proposal is brought forward to combine two departments into one and reassign some employees, issues worth studying are:

Cost:
- How much money will be saved by combining two departments?
- Are the savings permanent or will current expense levels return?
- What is the savings from reducing employee levels or higher efficiency of floor space?

Restrictions:
- Is it practical to combine these departments?
- What internal checks and balances might be compromised as a result?
- How much will the idea eliminate duplication? Are there other alternatives that cost less?

Time requirements:
- How long will it take to put this plan into effect?
- Will labor redundancies be absorbed through attrition and if so, over what period?
- What time span is involved in retraining employees?

3. *Support for proposals or initiatives.* In addition to deciding whether an idea is feasible or not, statistical analysis may help to make your case. When reviewing someone else's proposal, the same analysis may also be used to question whether the assumptions are rational and profitable. For example, in the case of combining departments, statistical study can be used to analyze current labor costs, estimate duplication of effort, and identify potential savings based on average wages. Additional savings may also come from higher efficiency in the use of floor space (for example, deferring the need for more rental expense due to expansion in the future).

   If a proposal is based on limited savings, a critical analysis of potential savings versus likely inefficiency is bolstered when studied financially. For example, if a limited amount of monthly savings is compared to the one-time cost of making the change, how many months will it take to break even? By that time, will expansion require breaking apart the combined department due to increased workload? A critical analysis of questions like this could make the decision less appealing due to only marginal potential benefits.

4. *Establishing standards.* Statistical analysis is perhaps at its greatest value when used to set standards. In a marketing forecast, you can support goals for a sales force based on volume of orders. In an expense budget, statistical reporting may be used to identify a reasonable level of expenses, providing you with the means for measuring variances throughout the year. In a production plant, analysis of shift defects and productivity can identify the most likely problem areas or weak links and point to methods to reduce problems.

Statistics are used in just about every aspect of analysis involving the 'p' numbers (production, personnel, and profits). Analysis is not simply a process undertaken by accountants and statisticians, but can also be applied to effectively reduce inefficiencies, avoid loss, bolster profitability, and make proposals convincing. All these outcomes are the result of setting standards and demonstrating how they can be enforced for a better future outcome.

One of the advantages to mastering math is that it improves your accuracy and makes you comfortable with financial information, even when your background does not include financial analysis. You vastly improve your ability to use math when you know a few easy but valuable shortcuts. Among the most useful of shortcuts is the estimate of an answer to a problem. If your answer is close to the estimate, it is probably right; but if it is off by millions, chances are good that the calculation includes a serious error. It fails the "test of reasonableness." The next chapter concludes with some useful math shortcuts in the basic functions as well as conversions between percentages, decimals and fractions.

Chapter 11
Incredible Math Shortcuts

As a manager, you are not doomed to using a calculator to figure out everything, or even to rely on spreadsheet programming to arrive at an answer. Many shortcuts will allow you to quickly and easily conquer the challenge that math presents.

This chapter lists many fast shortcuts in the first two basic functions: addition and subtraction. The purpose of this chapter is to provide an overview of some of the more practical applications of math shortcuts, and there is more advantage to learning these than simply mastering some tricks. Shortcuts also help you to:

- Estimate outcomes to avoid errors.
- Convert difficult numerical formats to more manageable ones before calculating answers.
- Master specific mathematical functions that have otherwise presented difficulties.

Addition

The first basic mathematical function is addition. In school you were taught to add using a well-known model, but it does not always guarantee fast results and is not necessarily the most efficient way to add.

Addition Shortcut # 1: Adding from the Left

This method involves first adding the far-right column (the '1's'), carrying over any digits beyond 9 and then repeating the process in the '10's' column. The first shortcut to learn is to abandon this system and add beginning on the *left* instead of on the right.

Example: A typical math exercise calls for addition of three large numbers:

$$8416$$
$$7715$$
$$+4499$$

DOI 10.1515/9781547400638-011

The traditional method of carrying digits over looks like this:

$$
\begin{array}{r}
{}^{1\,1\,2} \\
8416 \\
7715 \\
+4499 \\
\hline
20630
\end{array}
$$

Under this method, the process began at the far right where the three digits were added: 6+5+9=20. The '0' was placed beneath the far-right column and the '2' carried over as a remainder. Next, the '10's' column was added, including the remainder: 1+1+9+2=13. The '3' is listed in the answer and the '1' carried over to the '100's' column. It is then added as well: 4+7+4+1=16. This process is repeated until the answer of 20,630 has been developed. Now try the problem by starting at the left and carrying a series of balances:

$$
\begin{array}{r}
8416 \\
7715 \\
+4499 \\
\hline
19 \\
15 \\
11 \\
20 \\
\hline
1 \\
10 \\
6 \\
30 \\
\hline
20630
\end{array}
$$

The method takes up more space but does not involve more calculations; and no excess numerals need to be carried over as remainders. The steps go very quickly and result in improved accuracy as well.

Addition Shortcut # 2: Rounding up

Another shortcut requires that you round up numbers to be added and then subtract out the amount rounded. For example, here is a typical two-number addition problem:

$$37 + 49 = ?$$

This can be calculated easily enough but there is a quick shortcut as well. It requires you to round both numbers up to the next number divisible by 10:

$$40 + 50 = 90$$

The answer is quite easy; the final step is to subtract the value of the rounded-up numbers. The first number, 37, was rounded up three; and the second, 49, was rounded up 1. Subtract the total of 4 from the answer above:

$$40 + 50 = 90 - (3 + 1) = 86$$

Both steps—rounding up and subtracting the rounded value—can be done in your head and require less thinking than the traditional method. This also works for three or more numbers. For example:

$$47 + 58 + 66 = ?$$

or:

$$50 + 60 + 70 = 180 - (3 + 2 + 4) = 171$$

Addition Shortcut # 3: Fast Addition in Steps

In this shortcut, you calculate the additional values of a series of numbers, one at a time, adding by 10's as well as remainders. For example, when asked to add 37 and 49, you can quickly estimate with the process of steps. In your head, you quickly compute: ("37, 47, 57, 67, 77, 86." This involved adding the components of 49 in 10's: 10, 20, 30, 40, and 49, but starting with 37 as the base:

$$37 + 10 = 47 + 10 = 57 + 10 = 67 + 10 = 77 + 9 = 86$$

This seems like a lot of work just to add up two numbers, but it is the process you probably follow automatically without thinking about it. However, most people limit this to only two numbers. Now think about how you can apply the same shortcut to a larger series of two-digit numbers. For example:

$$47 + 58 + 66 = ?$$

or:

$$47 \quad 57, 67, 77, 87, 97, \mathbf{105}$$
$$105 \quad 115, 125, 135, 145, 155, 165, \mathbf{171}$$

The process of adding in your head by 10's and then attaching the remaining value works for any size list of two-digit numbers.

Addition Shortcut # 4: Fast Addition by Columns

The next shortcut breaks down the columns and adds them separately. Here is an example:

$$47 + 58 + 66 = ?$$

Or, first add the 10's columns and then add the 1's in sequence:

$$40 + 50 + 60 = 150 + 7 + 8 + 6 = 171$$

(counting steps)

$$90, 150, 157, 165, \mathbf{171}$$

This series of quick additions is easily done in your head and is a great exercise for improving rapid addition skills. Just looking at the columns of numbers, your fast mental addition takes you right to the total. "40, 90, 150, 157, 165, 171."

Subtraction

With subtraction, students are taught to use a precise and consistent formula. Like addition, subtraction can be made easier with a few easy shortcuts.

Subtraction Shortcut # 1: Add the Distances From 100

When you have to subtract two numbers, and one is above and the other is below 100 (or any other multiple of 100), you can use a fast shortcut to make subtraction easier. For example, you need to subtract 96 from 153 (this would also work with 196 and 253 because the distance between the two values is the same):

$$153 - 96 = 57$$
$$53 - 196 = 57$$

Another method is to calculate the distance of each number from 100, and then multiply the two results:

$$153 - 100 = 53$$
$$100 - 96 = 4$$
$$53 + 4 = 57$$

This series of steps can often be performed in your head and more rapidly than the traditional method, which involves moving values between columns:

$$
\begin{array}{r}
153 \\
-\ 96 \\
\end{array}
\quad = \quad
\begin{array}{r}
14^13 \\
-\ 96 \\
\hline
57 \\
\end{array}
$$

This traditional method requires moving the value of 10 from the 10's column and adding it to the 1's column. This changes '3' to '13' to take it larger than the 1's column below. This is a cumbersome series of steps that is difficult to teach and even more difficult to learn. It is much easier to rapidly isolate 53 and 4, and then add them together.

Subtraction Shortcut # 2: Alter Both Sides to Simplify

A second great shortcut, which also eliminates the "carrying the 10's" problem, is to simply increase both sides of the equation. Remember, when you change two sides equally, the answer remains the same. So, 36–29 is identical to the value of 37–30. However, the second version is easier to calculate:

$$36 - 29 = 7$$
$$37 - 30 = 7$$

The process works in reverse as well. For example, subtract 52 from 71 but *reduce* the values down to the closest round number in the lower value:

$$71 - 52 = 19$$
$$69 - 50 = 19$$

The adjustment of the value to be subtracted, either up or down to a round number, greatly simplifies the process, making it much easier to quickly do the calculation in your head.

Subtraction Shortcut # 3: Convert to Two Steps

When you have to move values from one column to another, it makes a problem more complicated than it needs to be. This is especially true when the subtraction task is presented in horizontal form. If you were taught to perform functions horizontally, the side-to-side format does not visually present you with a solution.

To overcome this, break down the subtraction problem into two easy steps. This works for any two 2-digit values. For example, you have to subtract 37 from 82:

$$82 - 37 = 45$$

To easily perform this in your head, first subtract the 10's column from the lower value; then subtract the remainder:

$$82 - 30 = 52$$
$$52 - 7 = 45$$

As with other methods, this can be rapidly performed in your head and without needing to cancel out the value of 10 and transfer it to the 1's column. The steps above make it easy to perform the process in your head, even for 3-digit values. For example:

$$191 - 126 = ?$$
$$191 - 120 = 71$$
$$71 - 6 = 65$$

This also works when the two values are not within the same 100's range:

$$215 - 162 = ?$$
$$215 - 160 = 55$$
$$55 - 2 = 53$$

This is somewhat more difficult to perform mentally because of the gap between the two values. Even so, removing the need to carry over values greatly simplifies the entire subtraction process.

Shortcuts come in handy for adding and subtracting but can be essential when dealing with the more complex formulas for multiplication and division. Multiplication has to begin with an explanation of a few logical rules.

Logical Rules

When you move beyond adding and subtracting and start multiplying numbers, the potential problems are expanded. Most people have some level of difficulty with multiplication, but using some simple shortcuts vastly simplifies the process.

First, a few logical multiplication rules need to be reviewed. A few rules of logic take a lot of mystery out of multiplication. These include:

Formula: Equivalents in Multiplication

$$\text{if } A = B$$
$$\text{and } A = C$$
$$\text{then } B = C$$

where: A, B, and C = any values

In other words, the equality of two related values dictates the third value as well. Another involves multiplication by the values of zero and one:

Formula: Multiplication by Zero and One

$$A \times 0 = 0$$
$$A \times 1 = A$$

where: A = any value

Check this on a spreadsheet:

A1   A
B1   =SUM(A1*0)
B2   =SUM(A1*1)

Summarizing these two universal truths, any value multiplied by zero is always equal to zero. And any number multiplied by one is always equal to the number itself (A). In a similar manner, another series of logical statements relate to multiplying by positive or negative values:

Formula: Multiplication of Positive and Negative Values

$$P_1 \times P_2 = P$$
$$P \times N = N$$
$$N_1 \times N_2 = P$$

where: P = positive value

N = negative value

On a spreadsheet:

A1 P_1 (first positive value)
A2 P_2 (second positive value
A3 N_1 (first negative value)
A4 N_2 (second negative value)
B1 =SUM(A1*A2)
C1 =SUM(A1*A3)
D1 =SUM(A3*A4)

A positive value multiplied by another positive *always* results in a positive answer. When a positive is multiplied by a negative, it is always a negative answer. And finally, when a negative value is multiplied by another negative value, it is always a positive answer.

Finally, rules apply when multiplying by even and odd numbers:

Formula: Multiplication by Even and Odd Values

$$E_1 \times E_2 = E$$
$$E \times O = O$$
$$O_1 \times O_2 = O$$

where: E: any even value

O: any odd value

In other words, all even numbers multiplied by any other even numbers always produce an even-numbered answer. However, in any operation involving any odd number, the answer will always be odd as well. This is true even if more than two values are in play.

Beyond the basic rules of multiplication are several useful shortcuts. These rules help clarify the process and enable you to perform many operations in your head.

Multiplication Shortcuts

When multiplying, the process is more complex than adding or subtracting. This added complexity causes some people to have greater trouble with multiplication than they should. For example, they find it easier to add 23, 23, and 23; but multiplying 23 by 3 poses a greater problem. This learned inhibition can be undone with the help of some useful multiplication shortcuts.

Multiplication Shortcut # 1: Squaring Numbers

One form of multiplication is squaring, or multiplying a number by itself:

Formula: Squaring a Number

$$n^2 = S$$

      where: n = any value
              S = square of the value

On a spreadsheet program:
    A1   =SUM(n*n)

When you square any number ending in 5, there is also a convenient shortcut. First, multiply the 10's digits by the next whole number. Then affix 25 to the result.

Formula: Square of Any Number Ending in 5

$$t \times (t + 1):25 = n^2$$

      where: t = 10's value of the number to be squared (not including 5)
             : = affix 25 to the product of t and t +1

To prove this on a spreadsheet:
    A1 = SUM(t*(t+1):25)

Example: Calculate the square of 25:

$$25^2 = ?$$
$$2 \times 3 = 6$$
$$6 \text{ affixed to } 25 = 6:25 \text{ or } 625$$

This works with any values. For example, to compute the square of 85:

$$85^2 = ?$$
$$8 \times 9 = 72$$
$$72 \text{ affixed to } 25 = 72{:}25, \text{ or } 7{,}225$$

Another method involves squaring of numbers ending in 1. In these cases, first drop the 1 from the value; next, find the square of the rounded number; and then add the original number to the newly calculated number.

Formula: Square of Any Number Ending in 1

$$(n - 1)^2 + ((n + (n - 1)) = n^2$$

where: n = number to be squared

To prove this on a spreadsheet:
    A1   =SUM(n*n)

Example: Find the square of 71:

$$71^2 = ?$$
$$71 - 1 = 70$$
$$70^2 = 4{,}900$$
$$70 + 71 = 141$$
$$4{,}900 + 141 = 5{,}041$$

In this example, 71 was converted down to 70. When these two values are added to the square of 70, the answer is correct for 71^2. It is amazing and easy, but it always works.

Another nice squaring shortcut works when you need to find the answer for a number ending in 9. In this case, you increase the number to be squared by 1, to a round number; then find its square. Then add together the original number and the revised number. Finally, subtract the total from the square calculated in the first step.

Formula: Square of Any Number Ending in 9

$$(n + 1)^2 - (n + (n + 1)) = n^2$$

where: n = number to be squared

To prove this on a spreadsheet:
    A1   =SUM(number+1)
    B1   =SUM(A1*A1)

C1   =SUM(number+A1)
C1   =SUM(B1-C1)

Example: Find the square of 79:

$$79^2 = ?$$
$$79 + 1 = 80$$
$$80^2 = 6,400$$
$$80 + 79 = 159$$
$$6,400 - 159 = 6241$$

Another squaring technique works for any two-digit number starting with the digit 5. To perform this shortcut, first add 25 to the 1's digit. Then square the one's digit (to two places) and affix that square to the answer in the first step. For example, to find the squares of 53, 55 and 57:

$$53^2 = ?$$
$$3 + 25 = 28$$
$$3^2 = 9$$
$$28:09 = 2,809$$
$$55^2 = ?$$
$$5 + 25 = 30$$
$$5^2 = 25$$
$$30:25 = 3,025$$
$$57^2 = ?$$
$$7 + 25 = 32$$
$$7^2 = 49$$
$$32:49 = 3,249$$

Multiplication Shortcut # 2: Dropping the Zeros

Whenever you have to multiply two numbers together and one (or both) contains an ending value of zero, the steps are simplified by dropping the zeros. The most basic version of this is multiplying two values like:

$$60 \times 7 = ?$$

Drop the zero and then add it back to the answer:

$$6 \times 7 = 42$$
$$42 \text{ converts to } 420$$

With only one zero involved, this is a fairly easy process. However, when both values contain zeroes, you need to add them both. For example:

$$60 \times 70 = ?$$

Drop the zeroes and then add both of them back to the answer:

$$6 \times 7 = 42$$
$$42 \; converts \; to \; 4,200$$

To make this easier to remember, add one zero to the answer for each zero dropped. This works for any extension. For example:

$$600 \times 70 = ?$$

Drop all of the zeroes and then add them back to the answer:

$$6 \times 7 = 42$$
$$42 \; converts \; to \; 42,000$$

Multiplication Shortcut # 3: Multiplying by 4

If you can multiply by two, you can also multiply by four. This involves two steps: First, double the number. Second, double again. For example

$$62 \times 4 = ?$$

The solution to perform in your head takes two steps, both doubling:

$$62 \times 2 = 124$$
$$124 \times 2 = 248$$

This process can be applied to any value multiplied by four. However, as you begin involving larger numbers, it becomes more of a challenge.

Multiplication Shortcut # 4: Multiplying by 5

The trick to multiplying by five is to remember that five is one-half of 10; and multiplying by 10 is easy. You simply add a zero. This problem is solved in two ways; you can either multiply by 10 and then divide by two; or first divide by two and then multiply by 10.

Method # 1:

$$22 \times 5 = ?$$
$$22 \times 10 = 220$$
$$220 \div 2 = 110$$

Method # 1:

$$22 \times 5 = ?$$
$$22 \div 2 = 11$$
$$11 \times 10 = 110$$

Formula: Multiplication by 5

$$(n \times 10) \div 2 = a$$

where: n = number ending in 5
a = answer

To prove on a spreadsheet:
A1   =sum((n*10)/2)

The problem is more complex when multiplying an odd number; dividing it by two as a first step results in a fraction or decimal value. For example, if you are multiplying 23 by 5, one-half of 23 is expressed as either 11½ or as 11.5. For this reason, the easiest solution is to first multiply by 10. Any multiplication by an even number *always* produces an even number as a result, making it easier to divide by 2:

$$23 \times 5 = ?$$
$$23 \times 10 = 230$$
$$230 \div 2 = 115$$

Multiplication Shortcut # 5: Multiplying by 11 or 12

One shortcut appears magical because it is so easy and so fast. When multiplying any two digits by 11, insert the sum of the multiplier's digits in between them and you have the answer. For example, multiply 27 by 11:

$$2 + 7 = 9$$
convert 27 to: 2 9 7

This works for any multiplier whose digits add up to 9 or less. When the digits add up to 10 or more, the one's digit of the answer is applied as above, and the extra digit is carried to the left. For example, if you multiply 67 by 11:

$$6 + 7 = 13$$
convert 67 to: 6 3 7
carry the 1: 7 3 7

A second method is to multiply the number by 10 and then add the original number to the total. To multiply 27 by 11 under this method:

$$27 \times 10 = 270$$
$$270 + 27 = 297$$

Another example, using 67:

$$67 \times 10 = 670$$
$$670 + 67 = 737$$

To multiply by 12, the same process is used, but the original number is added twice to the multiple by 10:

$$27 \times 10 = 270$$
$$270 + 27 + 27 = 324$$
$$67 \times 10 = 670$$
$$670 + 67 + 67 = 804$$

Multiplication Shortcut # 6: Multiplying by 9

The trick for multiplying by 11 has a similar application when dealing with 9 as the multiplier. The solution is to multiply the number by 10, and then subtract the original number from the total. For example, 27 x 9 is:

$$27 \times 10 = 270$$
$$270 - 27 = 243$$

Another example based on 67:

$$67 \times 10 = 670$$
$$670 - 67 = 603$$

Multiplication Shortcut # 7: Multiplying by 15

Multiplying by 15 is easy because it involves first multiplying by 10, or simply adding a zero; and then adding half of that total. For example, again using 27 and 67:

$$27 \times 10 = 270$$
$$270 \div 2 = 135$$
$$270 + 135 = 405$$
$$67 \times 10 = 670$$
$$670 \div 2 = 335$$
$$670 + 335 = 1{,}005$$

Multiplication Shortcut # 8: Multiplying Two Numbers 2 or 4 Digits Apart

Another impressive trick involves multiplying two numbers separated by two. To perform this process, square the number in between the two and then subtract 1 from the answer. For example:

$$25 \times 27 = ?$$
$$26^2 = 676$$
$$676 - 1 = 675$$

Another example:

$$65 \times 67 = ?$$
$$66^2 = 4{,}356$$
$$4{,}356 - 1 = 4{,}355$$

A similar process is used for multiplying numbers 4 digits apart. First square the number exactly halfway between the two; and then subtract 4. For example:

$$25 \times 29 = ?$$
$$27^2 - 729$$
$$729 - 4 = 725$$

Another example:

$$65 \times 69 = ?$$
$$672 = 4,489$$
$$4,489 - 4 = 4,485$$

Division Shortcuts

Performing division problems is even more complex than multiplication and more difficult to envision in your head. However, many great shortcuts eliminate many of the problems of division calculations.

Division Shortcut # 1: Fast Division by 4

This is a very straightforward shortcut. Just cut the number in half and cut the result in half again. For example, to divide 74 by 4:

$$74 \div 2 = 37$$
$$37 \div 2 = 18.5$$

If the decimal values are confusing, increase the number by 10 and then divide the final result by 10:

$$74 \times 10 = 740$$
$$740 \div 2 = 370$$
$$370 \div 2 = 185$$
$$185 \div 10 = 18.5$$

Division Shortcut # 2: Fast Division by 5 or 25

To quickly and easily divide by 5, first multiply the number by 2 and then move the decimal point one place to the left. For example, dividing 28 by 5:

$$28 \div 5 = ?$$
$$28 \times 2 = 56$$
$$56 \text{ converts to: } 5.6$$

Another example, dividing 412 by 5:

$$412 \div 5 = ?$$
$$412 \times 2 = 824$$
$$824 \text{ converts to: } 82.4$$

A similar process is used when you divide by 25. First multiply the number by 4 and then insert or move the decimal point two places to the left. For example:

$$230 \div 25 = ?$$
$$230 \times 4 = 920$$
$$920 \text{ converts to: } 9.2$$

Another example, dividing 470 by 25:

$$470 \div 25 = ?$$
$$470 \times 4 = 1,880$$
$$1,880 \text{ converts to: } 18.8$$

A similar process is used when dividing by 125 (multiplying by 8 and further multiples ending in 25. Just as squaring of a number contains special properties when ending in 5, division of numbers with the same features is easier as well.

The processes involving the whole numbers in the preceding sections work just as well with decimal-based numbers. The secret to rapidly performing these same short-cuts is to begin by ignoring the decimal points and reinserting them into the answer. Knowing how many "carries" have to be involved should be based on estimating the right answer. The more you practice, the more proficient you become.

The next chapter concludes the incredible shortcuts section by examining many processes for conversion (between decimal, percentage, and fraction), measurements, and time shortcuts (for example, how long it takes to double or triple a deposit based on a fixed interest rate).

Chapter 12
Incredible Conversion, Measurement and Time Shortcuts

Among the valuable math shortcuts managers are likely to need are those involving conversions between different numerical systems; measurements; and fast estimates of time and yield. This chapter takes a look at these important areas.

Conversion

When you confront a particular system, it is often desirable to convert it to another system. For example, it is easier to multiply decimal-based values than those expressed in fractions. Conversions among decimals, fractions, and percentages are inevitable in many business applications.

In an earlier chapter, the method for converting percentage to decimal was explained and expressed in two formats. First was the decimal shift:

$$r.0\% = 0.0r \; decimal$$

Second was the process of dividing a percentage by 100:

$$r \div 100 = D$$

In converting between systems, the purpose usually is to make it easier to perform functions. For example, you can rapidly multiply by 75 with the use of fractions, using one of two methods. First, divide by half, then add half of the result:

Formula: Multiplying by 75 Using Division

$$((V \div 2) \times 1.5) \times 100 = P$$

where: V = value
P = product

On a spreadsheet, enter the following:
A1 = V
B1 = 1.5
C1 = SUM(A1*B1)
D1 = C1*100

DOI 10.1515/9781547400638-012

Example: You want to multiple 126 by 75. To perform this, apply the shortcut:

$$((126 \div 2) \times 1.5) \times 100 = 9{,}450$$

A second method involves the use of a fraction. The value 75 is equal to ¾ of 100. So, using this converted value, multiplying by 75 can be expressed in another way:

Formula: Multiplying by 75 Using a Fraction

$$V \times \text{¾} \times 100 = P$$

> where: V = value
> P = product

On a spreadsheet, enter:
A1   V
B1   SUM(A1*3)/4
C1   =SUM(B1*100)

Example: To multiply 126 by 75 using the fraction ¾:

$$126 \times \text{¾} \times 100 = 9{,}450$$

Basic math review: To multiply a whole number by a fraction, first multiply the whole number by the numerator of the fraction; and divide the answer by the denominator. For example:

$$(126 \times 3) \div 4 = 94.5$$

Many functions are easier when converted from one system to another. For example, to convert fractions to decimals, divide the numerator by the denominator.

Formula: Fraction Conversion to Decimal

$$n \div d = a$$

> where: n = numerator
> d = denominator
> a = answer, decimal form

On a spreadsheet:

A1   n
B1   d
C1   =SUM(A1/B1)

Example: To convert the fraction ¾ to decimal form:

$$3 \div 4 = 0.75$$

This works with fractions of any size or complexity. To convert the fraction 32/49ths to decimal form:

$$32 \div 49 = 0.653$$

You may also want to convert percentage-form to fraction. For this, consider the percentage as the numerator and add a denominator to one place greater than the whole number of the percentage. If the percentage is a fractional number, convert to fraction by multiplying both top and bottom by 10 for each decimal place.

Example: To convert 35% to a fraction, express it is fractional form:

$$35 / 100$$

Next, convert this to the lowest common denominator. Since both sides of the fraction are divisible by 5:

numerator: $35 \div 5 = 7$
denominator $100 \div 5 = 20$
The value of 35% is equal to the fraction $7/20$.

If the fraction is less than a whole number, multiply the result by 10 for each decimal place. For example, 0.35% is converted to fractional form with these steps:

Express as a fraction $0.35 / 100$
Multiply each side by 100 (two decimal places)
numerator $0.35 \times 100 = 35$
denominator $100 \times 100 = 10,000$

The percentage 0.35 is equal to the fraction $35/10,000$ or $7/2000$. If the percentage is 3.5, conversion is:

Express as a fraction $3.5 / 100$
Multiply each side by 10 (one decimal place)

| numerator | $3.5 \times 10 = 35$ |
| denominator | $100 \times 10 = 1{,}000$ |

The fractional equivalent of 3.5% is $^{35}/_{1000}$ or $^{7}/_{200}$.
In working with fractions, methods of performing basic math functions are not as simple as that for whole numbers. To add fractions, the traditional method is to first find the common denominator; and to then add the numerators. This can be tedious and difficult for complex fractions; but there is a valuable shortcut.

Formula: Adding Fractions

$$(n_1 \times d_2) + (d_1 \times n_2) = n_a$$
$$d_1 \times n_2 = d_a$$

where: n_1 = numerator, first fraction
d_1 = denominator, first fraction
n_2 = numerator, first fraction
d_2 = numerator, second fraction
n_a = numerator, answer
d_a = denominator, answer

On a spreadsheet:
A1 n_1
A2 d_1
B1 n_2
B2 d_2
C1 =SUM(A1*B2)+(A2*B1)
D1 =SUM(A2*B2)

Example: You are adding the fractions $^{2}/_{3}$ and $^{3}/_{4}$ together. Applying the formula:

| numerator: | $(2 \times 4) + (3 \times 3) = 17$ |
| denominator: | $d_1 \times n_2 = 12$ |

The answer is $^{17}/_{12}$ which next can be converted to a whole number with a fraction. This is necessary whenever the numerator is larger than the denominator. To make this conversion, subtract a whole number equivalent from the fraction; and then add the whole number to the result:

subtract excess: $$^{17}/_{12} - {}^{12}/_{12} = {}^{5}/_{12}$$
add the whole number: $$1\,({}^{12}/_{12})_{+}\,{}^{5}/_{12} = 1{}^{5}/_{12}$$

To make this conversion work for whole number equivalents above '1', the process requires conversion to fractional form. For example, to convert the fraction ${}^{37}/_{12}$ to a whole number with a remaining fraction:

subtract excess: $$^{37}/_{12} - {}^{36}/_{12} = {}^{51}/_{12}$$
add the whole number: $$3\,({}^{36}/_{12}) + {}^{1}/_{12} = 36\,{}^{1}/_{12}$$

To subtract fractions, the same process is used; however, the cross-multiplied fractions are subtracted from one another.

Formula: Subtracting Fractions

$$(n_1 \times d_2) - (d_1 \times n_2) = n_a$$
$$d_1 \times n_2 = d_a$$

where: n_1 = numerator, first fraction
d_1 = denominator, first fraction
n_2 = numerator, second fraction
d_2 = numerator, second fraction
n_a = numerator, answer
d_a = denominator, answer

On a spreadsheet:
A1 n_1
A2 d_1
B1 n_2
B2 d_2
C1   =SUM(A2*B1)-SUM(A1*B2)
D1   =SUM(A2*B2)

Example: To subtract ${}^{1}/_{3}$ from ${}^{5}/_{8}$:

$$(5 \times 3) - ((8 \times 1) = 7$$
$$8 \times 3 = 24$$
$$answer: {}^{5}/_{8} - {}^{1}/_{3} = {}^{7}/_{24}$$

To prove this outcome, check using the traditional method, in which both fractions are converted to their lowest common denominator; and the numerators are subtracted:

$$\tfrac{5}{8} - \tfrac{1}{3} = \tfrac{15}{24} - \tfrac{8}{24} = \tfrac{7}{24}$$

These steps were:
1. Convert to lowest common denominator. Both sides of the fraction $\tfrac{5}{8}$ is multiplied by three to arrive at $\tfrac{15}{24}$. Next, both sides of the fraction $\tfrac{1}{3}$ are multiplied by 8 to arrive at $\tfrac{8}{24}$.
2. The numerators are subtracted. So, 15 minus 8 = 7, and the answer is $\tfrac{7}{24}$.

To multiply fractions, just multiply both numerators to find the top portion; and then multiply both denominators.

Formula: Multiplying Fractions

$$n_1 \times n_2 = n_a$$
$$d_1 \times d_2 = d_a$$

where: n_1 = numerator, first fraction
d_1 = denominator, first fraction
n_2 = numerator, second fraction
d_2 = numerator, second fraction
n_a = numerator, answer
d_a = denominator, answer

On a spreadsheet:
A1 n_1
A2 d_1
B1 n_2
B2 d_2
C1 =SUM(A1*B1)
D1 =SUM(A2*B2)

Example: To multiply $\tfrac{1}{3}$ by $\tfrac{5}{8}$:

$$1 \times 5 = 5$$
$$3 \times 8 = 24$$
$$answer: \tfrac{1}{3} \times \tfrac{5}{8} = \tfrac{5}{24}$$

To multiply mixed values—whole numbers with fractions—two methods can be used. First, convert the mixed number to a decimal equivalent. To do this, divide the numerator by the denominator and affix the decimal value to the whole number; repeat for the second value; and the multiply both values.

Formula: Multiplying Mixed Numbers with Conversion to Decimal Form

$$((n_1 \div d_1) + w_1) \times ((n_2 \div d_2) + w_2) = a$$

where: n_1 = numerator, first value's fraction
d_1 = denominator, first value's fraction
w_1 = whole value portion, first part
n_2 = numerator, second value's fraction
d_2 = denominator, second value's fraction
w_2 = whole value portion, second part
a = answer

On a spreadsheet:
A1 n_1
A2 d_1
A3 w_1
B1 n_2
B2 d_2
B3 w_2
C1 =SUM(A1/A2)+A3
C2 =SUM(B1/B2)+B3
D1 =SUM(C1*C2)

Example: You need to multiply 17 ¾ by 42 ³/₈. Applying the formula:

$$((3 \div 4) + 17) \times ((3 \div 8) + 42) = 752.15625$$

To divide fractions, the shortcut is to reverse the numerator and denominator of the first fraction, and then multiply by the second.

Formula: Dividing Fractions

$$d_1 \times n_2 = n_a$$
$$n_1 \times d_2 = d_a$$

where: n_1 = numerator, first fraction
d_1 = denominator, first fraction
n_2 = numerator, first fraction
d_2 = numerator, second fraction
n_a = numerator, answer
d_a = denominator, answer

On a spreadsheet:

    A1 n_1
    A2 d_1
    B1 n_2
    B2 d_2
    C1 =SUM(A2*B1)
    D1 =SUM(A1*B2)

Example: To divide $5/8$ by $1/3$:

$$8 \times 1 = 8$$
$$5 \times 3 = 15$$
$$answer: \; {}^5/_8 \div {}^1/_3 = {}^8/_{15}$$

The solutions to working with fractions are methodical and logical. Learning and applying the methods to simple fractions demonstrates how they can be applied with equal ease to even the most complex fractions.

Measurements

Managers may face situations in which measurements are necessary. This may involve any process from finding the area of a department as part of a proration within the annual budget, to calculating the area of a triangular piece of land owned by the company, in order to calculate one of many real estate versions of profitability.

    Many of the space measurements you will need rely on the calculated value of *pi*. This lower-case Greek letter is denoted in formulas with the symbol π, which is the calculated sum of the circumference of any circle, divided by its diameter. The size of the circle does not vary, because the calculation will always result in the same answer.

Formula: Pi

$$C \div D = \pi$$

    where: C = circumference of a circle
            D = diameter of a circle
            π = Pi

On a spreadsheet:

    A1 C
    B1 D
    C1 =SUM(A1/B1)

Example: The circumference of a circle is 223 inches. Its diameter is 70.983 inches. Pi is equal to:

$$223 \div 70.983 = 3.1416$$

Basic math review: The circumference of a circle is the distance around its entire area. Diameter is the distance from any point on the outside of the circle, through the exact middle and to the opposite side.

Formula: Circumference of a Circle

$$\pi \times D = C$$

      where: π = Pi
              D = diameter
              C = circumference

On a spreadsheet:
    A1 3.1416
    B1 D
    C1 =SUM(A1*B1)

Example: The diameter of a circle is 3 inches. Circumference is:

$$3.1416 \times 3 = 9.425 \text{ inches}$$

The most basic measurement is that of area for either a square or a rectangle. In both case, length is multiplied by width.

Formula: Area of a Square or Rectangle

$$L \times W = A$$

      where: L = length
              W = width
              A = area

On a spreadsheet:
    A1 L

B1   W
C1   =SUM(A1*B1)

Example: A square measures 14 by 14 feet. Area is:

$$14' \times 14' = 196'$$

Example: A rectangle measures 16 by 25 feet. The formula:

$$16' \times 25' = 400'$$

If the measurement also includes inches, as an area often does, the entire formula should be converted to inches and multiplied; and then converted back to feet.

Formula: Area with feet and inches

$$((F_1 \times 12) + I_1) \times ((F_2 \times 12) + I_2)) \div 144 = A$$

where: F_1 = feet, first measurement
I_1 = inches, first measurement
F_2 = feet, second measurement
I_2 = inches, second measurement
A = area

On a spreadsheet:
A1 F_1
A2 I_1
B1 F_2
B2 I_2
C1   =SUM(A1*12)+A2
C2   =SUM(B1*12)+B2
D1   =SUM(C1*C2)/144

Example: The rectangle you need to measure is 16 feet, 3 inches by 25 feet, 4 inches. Applying the formula:

$$((16 \times 12) + 3) \times ((25 \times 12) + 4)) \div 144 = 411\,^2/_3.$$

The fractional foot, $^2/_3$ can be converted to inches as well:

$$^2/_3 \times 12 = 8$$

The answer is that the area is 411'8".

To compute the area of a circle you need to first calculate the radius, which is one-half of the diameter. The diameter is a measurement from one side to the next through the center of the circle, so radius represents the distance from any outer point on the circle to its exact middle.

Formula: Radius

$$D \div 2 = R$$

> where: D = diameter
> R = radius

On a spreadsheet:
A1   D
B1   =SUM(A1/2)

To calculate area of a circle, multiply the square of the radius by pi.

Formula: Area of a Circle

$$R^2 \times \pi = A$$

> where: R = radius
> π = pi
> A = area of the circle

On a spreadsheet:
A1   R
B1   =SUM(A1*A1)*3.1416

Example: The radius of a circle is five inches. Applying the calculation:

$$5^2 \times 3.1416 = 78.54$$

The area of triangles can also be calculated quite easily. No matter what type of triangle is involved (right angle, acute angle or obtuse angle), the area is calculated in the same way.

Formula: Area of a Triangle

$$(b \times a) \div 2 = A$$

b = base
a = altitude
A = area

On a spreadsheet:
    A1   b
    B1   a
    C1=SUM(A1*B1)/2

Basic math review: The "base" of a triangle is the measurement of its bottom line. "Altitude" is the distance between the base and the highest point. In calculations involving triangles, lower-case letters are used for calculating area, and upper-case letters are used for finding angles.

Example: A triangle's base is 7 inches and its altitude is 4 inches. Area is calculated using the formula:

$$(7" \times 4") \div 2 = 14"$$

The calculation of area for odd shapes involves variations and combinations of the basic formulas for areas or rectangles and triangles. A somewhat more complex formula applies when you need to calculate the volume of a rectangular solid.

Formula: Volume of a Rectangular Solid

$$L \times W \times H = V$$

where: L = length
        W = width
        H = height
        V = volume

On a spreadsheet:
    A1   L
    B1   W
    C1   H
    D1   =SUM(A1*B1*C1)

Example: Your company is going to store some items during a corporate move. You are comparing the storage capacity of three different storage units. You need to know the volume of each. They measure 9 x 8 x 12, 10 x 12 x 8, and 12 x 14 x 10. To calculate the volume of each:

$$9 \times 8 \times 12 = 864 \ cubic \ feet$$
$$10 \times 12 \times 8 = 960 \ cubic \ feet$$
$$12 \times 14 \times 10 = 1{,}680 \ cubic \ feet$$

The volume of a cylinder, such as a rural storage building, is more complex because of its circular construction.

Formula: Volume of a Cylinder

$$R^2 \times \pi \times H = V$$

where: R = radius
π = pi
H = height
V = volume of a cylinder

On a spreadsheet:
A1   R
B1   =SUM(A1*A1)*3.1416
C1   height
D1= =SUM(B1*C1)

Example: A cylinder has a radius of 15 feet and is 25 feet high. The volume of this structure is:

$$15^2 \times 3.1416 \times 25 = 17{,}671.5 \ cubic \ feet$$

Time Shortcuts

The last variety of shortcuts involves calculations of how long it will take to double or triple a sum of money left on deposit. These are estimates based on an assumed rate of interest. The first is known as the Rule of 72.

Formula: Rule of 72

$$72 \div i = Y$$

where: i = interest rate
Y = years required to double the fund

On a spreadsheet:
    A1   i
    B1   =SUM(72/A1)

Example: Your company has set up a reserve for cash flow and has deposited a sum of $5,000 into stock that yields an annual dividend of 6%. How long will it take to double?

$$72 \div 6 = 12 \text{ years}$$

The Rule of 72 is a popular one; however, a slightly more accurate variation is called the Rule of 69. In this formula, the same steps are involved as for the Rule of 72; but the value of 0.35 is added to the answer.

Formula: Rule of 69

$$(69 \div i) + .35 = Y$$

where: i = interest rate number
Y = years required to double the fund

On a spreadsheet:
    A1   i
    B1   =SUM(69/A1)+.35

Based on the previous example:

$$(69 \div 6) + .35 = 11.85 \text{ years (about 11 years, 10 months)}$$

A third estimation is the Rule of 113, which quickly approximates the time required to triple a fund.

Formula: Rule of 113

$$113 \div i = Y$$

where: i = interest rate number
Y = years required to triple the fund

On a spreadsheet:
A1   i
B1   =SUM(113/A1)

Applying this to the previous example of $5,000 on deposit at 6%:

$$(113 \div 6) = 18.8\ years$$

These estimations are handy for comparing alternatives when money is going to be left on deposit. Like so many mathematical functions, any shortcut is useful if it saves time and improvers accuracy. A shortcut can be used to rapidly calculate exact answers or to find the approximate answer. For any manager requiring the use of math–and that includes virtually all managers–being able to plug in a shortcut improves confidence and makes any task easier.

The purpose of this book has been to display common calculations in a manner enabling every manager to quickly and easily arrive at accurate answers. Too often, mathematical processes are made overly complex. Simplicity invariably provides the same result but with less work and greater comprehension. Every manager is capable of mastering the underlying ideas represented by formulas, and every manager can improve their communication to subordinates, other managers and executives by being able to express mathematical results with absolute clarity.

Appendix A
Summary of Formulas

Accumulated value of a series of deposits:

$$D \times ((1 + R)^n - 1) \div R = A$$

where: D = periodic deposit amount
R = periodic interest rate
n = number of periods
A= accumulated value

Adding fractions:

$$(n_1 \times d_2) + (d_1 \times n_2) = n_a$$
$$d_1 \times n_2 = d_a$$

where: n_1= numerator, first fraction
d_1 = denominator, first fraction
n_2= numerator, first fraction
d_2 = numerator, second fraction
n_a = numerator, answer
d_a = denominator, answer

After-tax return:

$$O \times (1 - T) = A$$

where: O = operating (pre-tax) profit
T = combined federal and state tax rate (in decimal form)
A = after-tax profit

Amortization:

$$C \div M = A$$

where: C = total cost
M = months to amortize
A = amortization per month

DOI 10.1515/9781547400638-013

Annual compounding:

$$(i + 1)^x \times P = D$$

where: i = annual interest rate
x = number of years
P = principal deposited
D = total debt as of the number of years

Annual straight-line depreciation:

$$B \div Y = D$$

where: B = basis
Y = years in the recovery period
D = annual depreciation

Annualized return:

$$(R \div H) \times Y = A$$

where: R = return
H = holding period
Y = periods in one year (12)
A = annualized return

Area of a circle:

$$R^2 \times \pi = A$$

where: R = radius
π = pi
A = area of the circle

Area of a square or rectangle:

$$L \times W = A$$

where: L = length

W = width

A = area

Area of a triangle:

$$(b \times a) \div 2 = A$$

where: b = base

a = altitude

A = area

Area with feet and inches:

$$(F_1 \times 12) + I_1) \times (F_2 \times 12) + I_2) \div 144 = A$$

where: F_1 = feet, first measurement

I_1 = inches, first measurement

F_2 = feet, second measurement

I_2 = inches, second measurement

A = area

Balance sheet:

$$A = L + N$$

where: A = assets

L = liabilities

N = net worth

Breakeven return:

$$I \div (1 - T) = B$$

where: I = inflation rate

T = effective tax rate, including both federal and state (in decimal form)

B = breakeven return

Cash flow:

$$I + (N + L + A + S + O) - (L + A + S + D + O) = C$$

where: I = net income
N = non-cash expenses
L = loan transactions
A = capital asset transactions
S = legal settlements and judgments
O = other adjustments
D = dividends paid
C = cash flow

Cash income:

$$I + D = C$$

where: I = net income
D = depreciation expense
C = cash income

Cash-on-cash return:

$$C \div I = R$$

where: C = net cash flow per year
I = initial cash investment
R = cash-on-cash return

Circumference of a circle:

$$\pi \times D = C$$

where: π = Pi
D = diameter
C = circumference

Cost of merchandise:

$$B + M - E = C$$

where: B = beginning balance of inventory
M = merchandise purchased
E = ending balance of inventory
C = cost of merchandise

Current ratio:

$$A \div L = R$$

where: A = current assets
L = current liabilities
R = current ratio

Current yield on a bond:

$$N \div V = C$$

where: N = nominal yield
V = current value of the bond
C = current yield

Daily compounding:

$$(1 + (R \div i))^n \times P = C$$

where: R = stated annual interest rate

i = periodic interest rate (365 days)
n = number of periods to be compounded
P = principal
C = compounded value

Daily periodic rate (365 days):

$$R \div 365 = i$$

where: R = stated annual interest rate
i = periodic interest rate (365 days)

Days' sales outstanding:

$$R \div (S \div 365) = D$$

where: R = Accounts receivable balance
S = one year's sales on credit
D = days' sales outstanding

Debt coverage ratio:

$$I \div D = R$$

where: I = net income
D = debt service
R = debt coverage ratio

Debt to capitalization ratio

$$D \div T = R$$

where: D = long-term debt
T = total capitalization
R = debt to capitalization ratio

Declining balance (150%) depreciation:

$$(B - P) \div Y \times 150\% = D$$

where: B = basis
P = previous years' accumulated depreciation
Y = years in the recovery period
D = annual depreciation

Declining balance (200%) depreciation:

$$(B - P) \div Y \times 200\% = D$$

where: B = basis
P = previous years' accumulated depreciation
Y = years in the recovery period

D = annual depreciation

Degrees of a circle:

$$P \times 360 = D$$

where: P = percent of the total
D = degrees

Dispersion factor:

$$\sqrt{v} \div (V_1 + V_2 + \ldots V_n) \div n = D$$

where: $\sqrt{v}$ = variance
V_1 = first value in a field of values
V_2 = second value in a field of values
n = number of values in the field
D = Dispersion factor

Dividend yield:

$$D \div P = Y$$

where: D = dividend
P = price per share
Y = dividend yield

Dividing fractions:

$$d_1 \times n_2 = n_a$$
$$n_1 \times d_2 = d_a$$

where: n_1 = numerator, first fraction
d_1 = denominator, first fraction
n_2 = numerator, first fraction
d_2 = numerator, second fraction
n_a = numerator, answer
d_a = denominator, answer

Earnings per share (EPS):

$$E \div S = EPS$$

where: E = earnings
        S = number of common shares outstanding
      EPS = earnings per share

EBITDA:

$$N - (I + T + D + A) = E$$

where: N = net income
      I = interest expense
     T = taxes
     D = depreciation
     A = amortization
     E = EBITDA

Equivalents in multiplication:

           If $A = B$
         and $A = C$
        then $B = C$
    where: A, B and C = any values

Expense variance:

$$B - E = V$$

where: B = year-to-date budget
      E = year-to-date expense
      V = variance

Fixed asset turnover:

$$S \div (B + E) \div p = T$$

where: S = sales
      B = beginning fixed asset value
      E = ending fixed asset value

p = number of periods in the average

T = fixed asset turnover

Fraction conversion to decimal:

$$n \div d = a$$

where: n = numerator

d = denominator

a = answer, decimal form

Gross margin:

$$G \div R = M$$

where: G = gross profit

R = revenue

M = gross margin

Gross profit:

$$R - C = G$$

where: R = revenue

C = direct costs

G = gross profit

Half-year convention:

$$C \div 2 = D$$

where: C = calculated full-year depreciation

D = depreciation, first year

Home office depreciation:

$$B \times (I \div A) \div 27.5 \times (of \div tf) = D$$

where: B = basis (purchase price)

I = improvement value (assessed)
A = assessed value, total
of = office square feet
tf = total square feet
D = depreciation allowed

Income statement:

$$R - C = G$$
$$G - E = O$$
$$O - N = P$$
$$P - T = Z$$

where: R = revenue
C = direct costs
G = gross profit
E = expenses
O = operating profit
N = non-operating income or expense
P = pre-tax profit
T = income tax liability
Z = net after-tax profit

Inflation rate:

$$(C - P) \div P = I$$

where: C = current CPI index
P = past CPI index
I = rate of inflation, CPI

Interest coverage:

$$E \div I = C$$

where: E = EBITDA
I = interest expense
C = interest coverage

Inventory turnover:

$$C \div (B + E) \div p = T$$

where: C = cost of goods sold
B = beginning inventory
E = ending inventory
p = number of periods in the average
T = inventory turnover

Liability-to-asset ratio:

$$L \div A = R$$

where: L = total liabilities
A = total assets
R = liability-to-asset ratio

Loan amortization:

$$L \times (R \times P^n) \div ((P^n)\text{-}1) = A$$

where: L = original balance of the loan
R = periodic interest rate (annual rate divided by periods per year)
P = present value of 1
n = number of periods (usually months)
A = required payment per period

Mean absolute deviation:

$$((V_1 - A)^2 + (V_2 - A)^2 + \ldots (V_n - A)^2) \div n = D$$

where: V_1 = first value in the field
A = average of the field
V_2 = second value in the field
V_n = last value in the field of 'n'
n = number of values in the field
D = mean absolute deviation

Mid-month convention:

$$C \times (n \div 24) = D$$

where: C = calculated depreciation
n = monthly fraction
D = first-year depreciation

Mid-quarter convention:

1st quarter	$C \times (1.5 \div 12) \times 7 = D$
2nd quarter	$C \times (1.5 \div 12) \times 5 = D$
3rd quarter	$C \times (1.5 \div 12) \times 3 = D$
4th quarter	$C \times (1.5 \div 12) \times 1 = D$

where: C = calculated depreciation
D = first-year depreciation

Monthly compounding:

$$((i \div 12) + 1)^x \times P = D$$

where: i = annual interest rate
x = number of months
P = principal deposited
D = total debt as of the number of months

Monthly interest on a loan:

$$I = B \times (i \div p)$$

where: I = monthly interest
B = balance forward
i = annual interest rate
p = number of periods in the year (usually 12 months)

Monthly principal on a loan:

$$P = M - I$$

where: P = monthly principal
M = monthly payment
I = monthly interest

Monthly straight-line depreciation:

$$B \div (Y \times 12) = D$$

where: B = basis
Y = years in the recovery period
D = monthly depreciation

Multiplication by even and odd values:

$$E_1 \times E_2 = E$$
$$E \times O = O$$
$$O_1 \times O_2 = O$$

where: E: any even value
O: any odd value

Multiplication by zero and one:

$$A \times 0 = 0$$
$$A \times 1 = A$$

where: A = any value

Multiplication of positive and negative values:

$$P_1 \times P_2 = P$$
$$P \times N = N$$
$$N_1 \times N_2 = P$$

where: P = positive value
N = negative value

Multiplication by 5:

$$(n \times 10) \div 2 = a$$

where: n = number ending in 5
a = answer

Multiplying by 75 using a fraction:

$$V \times \tfrac{3}{4} \times 100 = P$$

where: V = value
P = product

Multiplying by 75 using division:

$$(V \div 2) \times 1.5 \times 100 = P$$

where: V = value
P = product

Multiplying fractions:

$$n_1 \times n_2 = n_a$$
$$d_1 \times d_2 = d_a$$

where: n_1 = numerator, first fraction
d_1 = denominator, first fraction
n_2 = numerator, second fraction
d_2 = numerator, second fraction
n_a = numerator, answer
d_a = denominator, answer

Multiplying mixed numbers with conversion to decimal form:

$$((n_1 \div d_1) + w_1) \times ((n_2 \div d_2) + w_2) = a$$

where: n_1 = numerator, first value's fraction
d_1 = denominator, first value's fraction
w_1 = whole value portion, first part

n_2 = numerator, second value's fraction
d_2 = denominator, second value's fraction
w_2 = whole value portion, second part
a = answer

Net after-tax profit:

$$P - T = Z$$

where: P = pre-tax profit
T = income tax liability
Z = net after-tax profit

Net return:

$$P \div R = N$$

where: P = net profit
R = revenue
N = net return

Net return on equity:

$$P \div (E - S) = N$$

where: P = net profit
E = equity (net worth)
S = Redeemable preferred stock
N = net return on equity

New balance forward on a loan:

$$N = B - P$$

where: = New monthly balance forward
B = balance forward
P = monthly principal

Operating profit:

$$G - E = O$$

where: G = gross profit
E = expenses
O = operating profit

Payback ratio:

$$I \div C = R$$

where: I = initial cash investment
C = net cash flow per year
R = payback ratio

Percent of expense variance:

$$V \div B = P$$

where: V = year-to-date variance (favorable or unfavorable)
B = year-to-date budget
P = percent of expense variance

Percent of the total:

$$V \div T = P$$

where: V = value
T = total
P = percent of the total

Percentage change:

$$(N - O) \div O = P$$

where: N = new base value
O = old base value
P = percentage change

Percentage Conversion to Decimal:

a) decimal shift: $r.0\% = 0.0r\ decimal$
b) divide by 100: $r \div 100 = D$

where: r = percentage rate
D = decimal equivalent

Percentage of revenue:

$$C \div R = P$$

where: C = income statement component
R = revenue
P = percentage

Periodic Rate:

$$R \div p = i$$

where R = nominal interest rate
p = number of periods
i = periodic interest rate

Pi:

$$C \div D = \pi$$

where: C = circumference of a circle
D = diameter of a circle
π = Pi

Present value of a single deposit:

$$D \div (1 + (i \div p))^n = V$$

where: i = annual interest rate
p = number of periods in the compounding method
n = periods until the deposit amount is needed
D = end-result deposit

V = amount needed to be deposited today

Present value per period:

$$W \times (1 \div (1 + (i \div p))^n \div (i \div p) = D$$

where: W = periodic withdrawal amounts
i = annual interest rate
p = number of periods in the compounding method
n = periods until the deposit amount is depleted
D = initial deposit required

Pre-tax net profit:

$$O +(-) N = P$$

where: O = operating profit
N = non-operating income or expense
(net income added, net expense deducted)
P = pre-tax profit

Price/earnings ratio:

$$P \div E = PE$$

where: P = price per share
E = EPS
PE = price/earnings ratio

Proof of proration:

$$P_a + P_b = V$$

where: P_a = prorated value of a
P_b = prorated value of b
V = value to be prorated

Proration:

$$V(a / (a + b)) = Pa$$
$$V(b/ (a + b)) = Pb$$

where: V = value to be prorated
a = proration base factor a
b = proration base factor b
P_a = prorated value of a
P_b = prorated value of b

Quarterly compounding:

$$((i \div 4) + 1)^x \times P = D$$

where: i = annual interest rate
x = number of quarterly periods
P = principal deposited
D = total debt as of the number of periods

Quick assets ratio:

$$(A - I) \div L = R$$

where: A = current assets
I = inventory
L = current liabilities
R = current ratio

Radius:

$$D \div 2 = R$$

where: D = diameter
R = radius

Remaining balance percentage on a loan:

$$R = N \div L$$

where: R = remaining balance percentage
N = New monthly balance forward
L = original loan amount

Return on cash invested:

$$(S - P) \div I = R$$

where: S = sales price
P = purchase price
I = cash invested
R = return on cash invested

Return on equity:

$$P \div E = N$$

where: P = net profit
E = equity (net worth)
N = return on equity

Return on net investment:

$$(S - P - C) \div I = R$$

where: S = sales price
P = purchase price
C = costs
I = cash invested
R = return on cash invested

Return on purchase price:

$$(S - P) \div P = R$$

where: S = sales price
P = purchase price
R = return on purchase price

Rule of 113:

$$113 \div i = Y$$

where: i= interest rate
Y = years required to triple the fund

Rule of 69:

$$(69 \div i) + .35 = Y$$

where: i= interest rate
Y = years required to double the fund

Rule of 72:

$$72 \div i = Y$$

where: i= interest rate
Y = years required to double the fund

Semiannual compounding:

$$((i \div 2) + 1)^x \times P = D$$

where: i = annual interest rate
x = number of semiannual periods
P = principal deposited
D = total debt as of the number of periods

Simple average:

$$(V_1 + V_2 + \ldots V_n) \div n = A$$

where: V = value
1, 2 = field number
n = last number in the field
A = average

Simple interest:

$$P \times R = I$$

where: P = principal
R = interest rate
I = interest

Sinking fund payments:

$$D \times ((i \div p) \div (1 + (i \div p))^n - 1) = V$$

where: D = target deposit
i = annual interest rate
p = number of periods in the compounding method
n = periods until the deposit amount is needed
V = amount of periodic deposits required

Square of any number ending in 1:

$$(n - 1)^2 + (n + (n - 1)) = n^2$$

where: n = number to be squared

Square of any number ending in 5:

$$t \times (t + 1):25 = n^2$$

where: t = 10's value of the number to be squared (not including 5)
: = affix 25 to the product of t and t +1

Square of any number ending in 9:

$$(n + 1)^2 - ((n + (n + 1)) = n^2$$

where: n = number to be squared

Squaring a number:

$$n^2 = S$$

where: n = any value
S = square of the value

Standard deviation:

$$\sqrt{V}$$

The square route of the variance.

Subtracting fractions:

$$(n_1 \times d_2) - (d_1 \times n_2) = n_a$$
$$d_1 \times n_2 = d_a$$

where: n_1 = numerator, first fraction
d_1 = denominator, first fraction
n_2 = numerator, first fraction
d_2 = numerator, second fraction
n_a = numerator, answer
d_a = denominator, answer

Variance:

$$((V_1)^2 + (V_2)^2 + ... (V_n)^2) \div (n\text{-}1) - (V_1 + V_2 + ... V_n)^2 = VR$$

where: V_1 = first value in the field
V_2 = second value in the field
V_n = last value in the field of 'n' values
N = number of values in the field
VR = variance

Volume of a cylinder:

$$R^2 \times \pi \times H = V$$

where: R = radius

π = pi
H = height
V = volume of a cylinder

Volume of a rectangular solid:

$$L \times W \times H = V$$

where: L = length
W = width
H = height
V = volume

Weighted average:

$$I_1 \times (L_1 \div L_t) + I_2 \times (L_2 \div L_t) = W$$

where: I_1 = interest rate, loan '1'
L_1= borrowed amount, loan '1'
L_t = total of amounts borrowed
I_2 = interest rate, loan '2'
L_2 = borrowed amount, loan '2'
W = weighted average

Year-to-date budget:

$$C + Y = B$$

where: C = current-month budget
Y = prior year-to-date budget
E = year-to-date budget

Year-to-date expense:

$$C + Y = E$$

where: C = current-month expense
Y = prior year-to-date expense
E = year-to-date expense

Appendix B
Summary of Excel Programs

Accumulated value of a series of deposits:

A1 Deposits
B1 =SUM(1+Rate)
C1 =B1
B2 =B1
C2 =SUM(C1*B2)

Copy B and C, row 3
Paste to subsequent B and C rows

D36 =SUM(C36-1)/(C1-1)
E36 =SUM(D36*A1)

Adding fractions:

A1 n_1
A2 d_1
B1 n_2
B2 d_2
C1 =SUM(A1*B2)+(A2*B1)
D1 =SUM(A2*B2)

After-tax return:

A1 O
B1 T (in decimal form)
C1 =SUM(1-B1)*A1

Amortization:

A1 C
B1 M
C1 =SUM(A1/B1)

Annual compounding:

A1	Annual interest rate plus 1	=SUM(i+1)
B1	Principal amount	
C1	Accumulated amount	= SUM(A1*B1)
A2		=A1
B2		=C1

DOI 10.1515/9781547400638-014

To perform this Excel calculation, do the following:

3. Copy C1
4. Paste to C2
5. Copy A2, B2, and C2
6. Paste to rows 3, columns 1, 2, and 3
7. Repeat paste for each row

Annual straight-line depreciation:

A1   B
B1   Y
C1   =SUM(A1/B1)

Annualized return:

A1   R
B1   H
C1   =SUM(A1/B1)*12

Area of a circle:

A1   R
B1   =SUM(A1*A1)*3.1416

Area of a square or rectangle:

A1   L
B1   W
C1   =SUM(A1*B1)

Area of a triangle:

A1   b
B1   a
C1=SUM(A1*B1)/2

Area with feet and inches:

A1 F_1
A2 I_1
B1 F_2
B2 I_2
C1   =SUM(A1*12)+A2
C2   =SUM(B1*12)+B2
D1   =SUM(C1*C2)/144

Balance sheet:

Spreadsheet fields:
- A1 Assets
- B1 Liabilities
- C1 Net worth
- D1 =SUM(B1+C1) (should match cell A1)

Breakeven return:
- A1 Inflation rate
- B1 Effective tax rate
- C1 =SUM(A1/(1-B1))

Cash flow:
- A1 Net income
- B1 Non-cash expenses
- B2 Loan transactions
- B3 Capital asset transactions
- B4 Legal settlements and judgments
- B5 Other adjustments
- B6 =SUM(B1:B5)
- C1 Loan transactions
- C2 Capital asset transactions
- C3 Legal settlements and judgments
- C4 Dividends paid
- C5 Other adjustments
- C6 =SUM(C1:C5)
- D6 =SUM(A1+B6-C6)

Cash income:
- A1 Net income
- B1 Depreciation expense
- C1 =SUM(A1+B1)

Cash-on-cash return:
- A1 Net cash flow per year
- B1 Initial cash investment
- C1 =SUM(A1/B1)

Circumference of a circle:
- A1   3.1416
- B1   Diameter
- C1   =SUM(A1*B1)

Cost of merchandise:
- A1   Beginning balance of inventory
- A2   Merchandise purchased
- A3   Ending balance of inventory
- A4   =SUM(A1+A2-A3)

Current ratio:
- A1   Assets
- B1   Liabilities
- C1   =SUM(A1/B1)

Current yield on a bond:
- A1   Nominal yield
- B1   Current value of the bond
- C1   =SUM(A1/B1)

Daily compounding:
- A1   Annual interest rate divided by 365 = daily rate, plus 1   =SUM(i/365) + 1
- B1   Principal amount
- C1   Accumulated amount   =SUM (A1*B1)
- A2   =A1
- B2   =C1

To perform this Excel calculation, do the following:
1. Copy C1
2. Paste it to C2, so that C1 and C2 are the same
3. Copy the row A2, B2, and C2
4. Paste to row 3, columns A, B, and C
5. Repeat paste for each row

Daily periodic rate (365 days):

$$R \div 365 = i$$

where: R = stated annual interest rate

         i = periodic interest rate (365 days)

Spreadsheet fields:

A1   Stated annual interest rate (in decimal form)

B1   365

C1 =SUM(A1/B1)

Days' sales outstanding:

A1   Accounts receivable balance

B1   One year's sales on credit

C1 =SUM(A1)/(B1/365)

Debt coverage ratio:

A1   Net income

B1   Debt service

C1 =SUM(A1/B1)

Debt to capitalization ratio:

A1   Long-term debt

B1   Total capitalization

C1 =SUM(A1/B1)

Declining balance (150%) depreciation:

A1   Basis

B1   =SUM((A1/5)*1.5)

C1   =SUM(A1-B1)

A2   =C1

B2, C2   Copy B1 and C1; paste to B2 and C2

row 3   Copy all cells, row 2; paste to row 3

Declining balance (200%) depreciation:

A1   Basis

B1   =SUM((A1/5)*2)

C1   =SUM(A1-B1)

A2   =C1

B2, C2   Copy B1 and C1; paste to B2 and C2

row 3   Copy all cells, row 2; paste to row 3

Degrees of a circle:
    A1   Percent of the total
    B1 =SUM(A1*360)

Dispersion factor:
    A1   =SQRT(v)
    B1 through B_n Values
    C1   =SUM($B1:B_n$)/ A
    D1   =SUM(A1/C1)

Dividend yield:
        A1   Dividend
        B1   Price per share
        C1   =SUM(A1/B1)

Dividing fractions:

$$d_1 \times n^2 = n^a$$
$$n_1 \times d_2 = d_a$$

        where: n_1 = numerator, first fraction
                   d_1 = denominator, first fraction
                   n_2 = numerator, second fraction
                   d_2 = numerator, second fraction
                   n_a = numerator, answer
                   d_a = denominator, answer

Spreadsheet fields:
    A1 Numerator, first fraction
    A2 Denominator, first fraction
    B1 Numerator, second fraction
    B2 Denominator, second fraction
    C1 =SUM(A2*B1)
    D1 =SUM(A1*B2)

Earnings per share (EPS):
    A1 Earnings
    B1 Number of common shares outstanding

C1   =SUM(A1/B1)

EBITDA:

A1   Net income
A2   Interest expense
A3   Taxes
A4   Depreciation
A5   Amortization
A6   =SUM(A1:A5)

Equivalents in multiplication:

A1   Value of A
B1   Value of B
C1   Value of C
A2   =SUM(A1*B1)
B2   =SUM(A1*C1)
C2   =SUM(B1*C1)

Expense variance:

A1   Year-to-date budget
B1   Year-to-date expense
C1   =SUM(A1-B1)

Fixed asset turnover:

A1   S
A2   B
A3   E
A4   Ending inventory
A5   Number of periods in the average
A6   =SUM(A1/((A2+A3)/p))

Fraction conversion to decimal:

A1   Numerator
B1   Denominator
C1   =SUM(A1/B1)

Gross margin:

A1   Gross profit
B1   Revenue

C1 =SUM(A1/B1)

Gross profit:
A1 Revenue
A2 Direct costs
A3 =SUM(A1-A2)

Half-year convention:
A1 Calculated full-year depreciation
B1 =(SUM(A1/2)

Home office depreciation:
A1 B
A2 I
A3 A
B1 of
B2 tf
C1 =SUM((A1*(A2/A3)))
C2 =SUM(C1/27.5) (note: 27.5 is the number of years depreciation can be claimed)
C3 =SUM((C2*(B1/B2)))

Income statement:
A1 Revenue
A2 Direct costs
A3 =SUM(A1-A2)
A4 Expenses
A5 =SUM(A3-A4)
A6 Non-operating income or expense
A7 =SUM(A5-A6)
A8 Income tax liability
A9 =SUM(A7-A8)

Inflation rate:
A1 Current CPI index
B1 Past CPI index
C1 =SUM(A1-B1)/B1

Interest coverage:
A1 EBITDA

B1   Interest expense
C1   =SUM(A1/B1)

Inventory turnover:
  A1   Cost of goods sold
  B1   Beginning inventory
  B2   Ending inventory
  C1   Number of periods in the average
  C2   =SUM(A1/((B1+B2)/p))

Liability-to-asset ratio:
  A1   Total liabilities
  B1   Total assets
  C1   =SUM(A1/B1)

Loan amortization:
  A1   =SUM(1+(i/p))
  B1   =A1
  A2   =A1
  B2   =SUM(A2*B1)
            Copy row 2, columns A and B
            Paste to subsequent rows
  C360   =SUM(A360-1)*B360 (assuming a 30-year mortgage term, or
            a total of 360 months)
  D360   =SUM((C360/(B360-1))*L

Loan payments and balances:

		B
A1	Loan amount ($10,000 in the example used)	
B1	Monthly interest (7.5% per year, compounded monthly)	=SUM(A1*(i/p))
C1	Monthly principal ($69.22 for 30 years in the example)	=SUM(M-B1)
D1	New balance forward (previous less principal)	=SUM(A1-C1)
E1	Remaining balance percentage	=SUM(D1/A1)
A2		=D1

To perform this Excel calculation, do the following:
1. Copy B1, C1, D1, E1
2. Paste to B2, C2, D2, E2
3. Copy columns A through E, row 2
4. Paste to all subsequent cells (for as long as calculation is to be performed, such as 360 for a 30-year loan)

Mean absolute deviation:

A1 through A$_n$	V (one value per cell through cell 'n')
B1	Average of the field
	Copy B1 and paste for all 'B' cells
C1 through C$_n$	=SUM(A1-B1)*(A1-B1)
	Copy C1 and paste for all 'C' cells
D1	= SUM(C1:Cn)/n

Median:

A1:An	F
B1	=MEDIAN(A1:An)

Mid-month convention:

A1	Calculated depreciation
B1	=SUM(n/24)
C1	=(SUM(A1*B1)

Mid-quarter convention:

A1	Calculated depreciation
B1	=SUM((1.5/12)*x
C1	=(SUM(A1*B1)

Monthly compounding:

A1	Annual interest rate plus 1	=SUM(i/12) + 1
B1	Principal amount	
C1	Accumulated amount	=SUM(A1*B1)
A2		=A1
B2		=C1

To perform this Excel calculation, do the following:
1. Copy C1
2. Paste to C2
3. Copy A2, B2, and C2
4. Paste to rows 3, columns 1, 2, and 3

Monthly interest on a loan:

$$I = B \times (i \div p)$$

where: I = monthly interest

B = balance forward

i = annual interest rate

p = number of periods in the year (usually 12 months)

Spreadsheet fields:

A1   Balance forward

B1   =SUM(interest rate/12)

C1   =SUM(A1*B1)

Monthly principal on a loan:

A1   Monthly payment

B1   Monthly interest

C1   =SUM(A1-B1)

Monthly straight-line depreciation:

A1   Basis

B1   =SUM(years of recovery*12)

C1   =SUM(A1/B1)

Multiplication by even and odd values:

A1 E_1 (first even value)

A2 E_2 (second even value)

A3 O_1 (first odd value)

A4 O_2 (second odd value)

B1   =SUM(A1*A2)

C1   =SUM(A1*A3)

D1   =SUM(A3*A4)

Multiplication by zero and one:

A1   Any value

B1   =SUM(A1*0)

B2   =SUM(A1*1)

Multiplication of positive and negative values:

A1 P_1 (first positive value)

A2 P_2 (second positive value

A3 N_1 (first negative value)

A4 N_2 (second negative value)

B1	=SUM(A1*A2)
C1	=SUM(A1*A3)
D1	=SUM(A3*A4)

Multiplication by 5:

A1	=sum((number ending in 5*10)/2)

Multiplying by 75 using a fraction:

A1	Value
B1	SUM(A1*3)/4
C1	=SUM(B1*100)

Multiplying by 75 using division:

A1 = V
B1 = 1.5
C1 = SUM(A1*B1)
D1 = C1*100

Multiplying fractions:

A1	Numerator, first fraction
A2	Denominator, first fraction
B1	Numerator, second fraction
B2	Numerator, second fraction
C1	=SUM(A1*B1)
D1	=SUM(A2*B2)

Multiplying mixed numbers with conversion to decimal form:

A1	Numerator, first value's fraction
A2	Denominator, first value's fraction
A3	Whole value portion, first part
B1	Numerator, second value's fraction
B2	Denominator, second value's fraction
B3	Whole value portion, second part
C1	=SUM(A1/A2)+A3
C2	=SUM(B1/B2)+B3
D1	=SUM(C1*C2)

Net after-tax profit:

A7	Pre-tax profit

A8 Income tax liability
A9 =SUM(A7-A8)

Net return:
A1 Net profit
B1 Revenue
C1 =SUM(A1/B1)

Net return on equity:
A1 Net profit
B1 Equity
C1 Redeemable preferred stock
D1 =SUM(A1/(B1-C1))

New balance forward on a loan:
A1 Balance forward
B1 Monthly principal
C1 =SUM(A1-B1)

Operating profit:
A3 Gross profit
A4 Expenses
A5 =SUM(A3-A4)

Payback ratio:
A1 Initial cash investment
B1 Net cash flow per year
C1 =SUM(A1/B1)

Percent of expense variance:

$$V \div B = P$$

where: V = year-to-date variance (favorable or unfavorable)
B = year-to-date budget
P = percent of expense variance

Spreadsheet fields:
A1   Year-to-date variance (favorable or unfavorable)

B1   Year-to-date budget
C1   =SUM(A1/B1)

Percent of the total:
A1   T
         Copy A1
         Paste to A2, A3, A4, etc.
B1   V(1)
B2   V(2)
B3   V(3)
B4   V(4)
C1   =Sum(B1/A1)
         Copy C1
         Paste to C2, C3, C4, etc.

Percentage change:
A1   New base value
B1   Old base value
C1   =SUM((A1-B1)/B1)

Percentage Conversion to Decimal:
A1   Percentage rate
B1   =SUM(A1/100)

Percentage of revenue:
A1   Income statement component
A2   Revenue
A3   =SUM(A1/A2)

Periodic Rate:
A1   Nominal interest rate
B1   Number of periods
C1   =SUM(A1/B1)

Pi:
A1   Circumference of a circle
B1   Diameter of a circle
C1   =SUM(A1/B1)

Present value of a single deposit:

A1 =SUM(1+(interest rate/number of periods))
B1 =A1
A2 =A1
B2 =SUM(A2*B1)
    Copy row 2, columns A and B
    Paste to subsequent rows
C60 =SUM(1/B60) (assuming 60 months total)
D60 =SUM(C60*D)

Present value per period:

A1 =SUM(1+(annual interest rate/number of periods))
B1 =A1
A2 =A1
B2 =SUM(A2*B1)
    Copy row 2, columns A and B
    Paste to subsequent rows
    C60   =SUM(1-(1/B60)) (assuming 60 months total)
D60 =SUM(C60/(A1-1))*W

Pre-tax net profit:

A5 Operating profit
A6 Non-operating income or expense
A7 =SUM(A5-A6)

Price/earnings ratio:

A1 Price per share
B1 Earnings per share
C1 =SUM(A1/B1)

Proof of proration:

F1 Pa
G1 Pb
H1 =SUM(F1+G1)

Proration:

A1 Value to be prorated
B1 Proration base factor a
C1 Proration base factor b

D1   =SUM(a/(a+b))
E1   =SUM(b/(a+b))
F1   =SUM(A1*D1)
G1   =SUM(A1*E1)

Quarterly compounding:

A1	Annual interest rate plus 1	=SUM(i/4) + 1
B1	Principal amount	
C1	Accumulated amount	= SUM(A1*B1)
A2		=A1
B2		=C1

To perform this Excel calculation, do the following:
1. Copy C1
2. Paste to C2
3. Copy A2, B2, and C2
4. Paste to rows 3, columns 1, 2, and 3

Quick assets ratio:

A1   Current assets
B1   Inventory
C1   Current liabilities
C1   =SUM(A1-B1)/C1

Radius:

A1   Diameter
B1   =SUM(A1/2)

Remaining balance percentage on a loan:

A1   New monthly balance forward
B1   Original loan amount
C1   =SUM(A1/B1)

Return on cash invested:

A1   Sales price
B1   Purchase price
C1   Cash invested
D1   =SUM(A1-B1)/C1

Return on equity:
- A1 Net profit
- B1 Equity
- C1 =SUM(A1/B1)

Return on net investment:
- A1 Sales price
- B1 Purchase price
- C1 Costs
- D1 Cash invested
- E1 =SUM(A1-B1-C1)/D1

Return on purchase price:
- A1 Sales price
- B1 Purchase price
- C1 =SUM(A1-B1)/B1

Rule of 113:
- A1 Interest rate
- B1 =SUM(113/A1)

Rule of 69:
- A1 Interest rate
- B1 =SUM(69/A1)+.35

Rule of 72:
- A1 Interest rate
- B1 =SUM(72/A1)

Semiannual compounding:

A1	Annual interest rate plus 1	=SUM(i/2)+1
B1	Principal amount	
C1	Accumulated amount	=SUM(A1*B1)
A2		=A1
B2		=C1

To perform this Excel calculation, do the following:
1. Copy C1
2. Paste to C2

3.   Copy A2, B2, and C2
4.   Paste to rows 3, columns 1, 2, and 3
5.   Repeat paste for each row

Simple average:

A1	Value 1
A2	Value 2
Final cell (A12)	Value (final)
B12	=SUM(A1:A12)/12

Simple interest:
A1   Principal
B1   Interest rate
C1   =SUM(A1*B1)

Sinking fund payments:
A1   =SUM(1+(annual interest rate/number of periods))
B1   =A1
A2   =A1
B2   =SUM(A2*B1)
      Copy row 2, columns A and B
      Paste to subsequent rows
C60 =SUM(A1-1)/(B60-1)
D60 =SUM(C60*D)

Square of any number ending in 1:
A1   =SUM(number*number)

Square of any number ending in 5:

$$t \times (t + 1):25 = n^2$$

where: t = 10's value of the number to be squared
: = affix

Spreadsheet fields:
A1   = SUM(t*(t+1):25)

Square of any number ending in 9:
    A1   =SUM(number+1)
    B1   =SUM(A1*A1)
    C1   =SUM(number+A1)
    C1   =SUM(B1-C1)

Squaring a number:
    A1   =SUM(n*n)

Standard deviation:
    A1   SQRT(value)

Subtracting fractions:
    A1   Numerator, first fraction
    A2   Denominator, first fraction
    B1   Numerator, second fraction
    B2   Numerator, second fraction
    C1   = SUM(A2*B1)-SUM(A1*B2)
    D1   =SUM(A2*B2)

Variance:
    A1 through A_n Values
    B1   =SUM(A1:A_n)/A
    C1   =SUM(A1-B1)*(A1-B1)
                                Copy C1 and paste to remaining 'C' cells
    D1   =SUM(C1:C_n)/n-1

Volume of a cylinder:
    A1   R
    B1   =SUM(A1*A1)*3.1416
    C1   Height
    D1= =SUM(B1*C1)

Volume of a rectangular solid:
    A1   Length
    B1   Width
    C1   Height
    D1   =SUM(A1*B1*C1)

Weighted average:
    A1 Borrowed amount, loan '1'
    B1 Borrowed amount, loan '2'
    C1 =SUM(A1+B1)
    A2 Interest rate, loan '1'
    B2 = interest rate, loan '2'
    A3 =SUM(A2*(A1/C1))
    B3 =SUM(B2*(B1/C1))
    C3 =SUM(A3+B3)

Year-to-date budget:
    A1 Current-month budget
    B1 Prior year-to-date budget
    C1 =SUM(A1+B1)

Year-to-date expense:
    A1 Current-month expense
    B1 Prior year-to-date expense
    C1 =SUM(A1+B1)

Glossary

Accelerated depreciation a form of depreciation allowing more write-off in the early years and less in later years; a faster rate of depreciation than under the straight-line method.

Accumulated depreciation a reduction of the fixed asset classification on the balance sheet, representing all depreciation claimed over the period that assets are owned.

Accumulated value of a series of deposits the ending value of a fund based on the amount of periodic deposit, interest rate, compounding method, and total period.

Acid test alternative name for the *quick assets ratio*.

After-tax income the dollar amount of net income after all deductions have been taken, including provisions for income taxes.

After-tax return the rate of return of the net income after all deductions, calculated by multiplying the operating (pre-tax) profit by the rate of after-tax income (the net after deducting the rate of federal and state taxation).

Amortization an expense similar to depreciation, used to expense intangible assets or to write off prepaid assets into the proper accounting year.

Annual compounding a method of compounding in which the calculation of interest is performed only once per year.

Annual percentage rate (APR) the overall interest based on calculation of the periodic rate compounded for a full year, plus applicable expenses or charges associated with a loan.

Annualized return a return expressed as if the holding period were exactly one full year, useful for comparing investments held for different periods.

Area of a circle the area, calculated by multiplying radius^2 by pi.

Area of a square or rectangle the length multiplied by the width.

Area of a triangle the base (size of the bottom length) multiplied by the altitude (height from base to highest point), the sum of which is then divided by two.

DOI 10.1515/9781547400638-015

Assets properties owned by companies or individual, contrasted with liabilities (debts) owned. The net difference between assets and liabilities is the equity or net worth.

Assumption (budgeting) the underlying facts taken into account in preparation of itemized budget levels for each expense, such as number of employees, square feet of departments, or contractual obligations.

Balance sheet a financial statement prepared as of a specified date, usually the end of a fiscal quarter or year. It contains the ending balances of all asset, liability and net worth accounts. The sum of all assets is equal to the sum of all liabilities plus net worth accounts.

Book value per share the value of a corporation on a per-share basis. The total book value (assets minus liabilities) is divided by the number of common shares of stock outstanding.

Breakeven return the rate of return needed to break even after allowing for both inflation and taxes. To calculate, the estimated inflation rate is divided by the after-tax earnings rate, and the answer is expressed as a percentage.

Budget an annual or semiannual estimate of future expenses, based on assumptions and expected changes within the organization; used as a means for monitoring internal controls and maintaining expected levels of profitability.

Capital assets properties owned by a company and subject to depreciation over several years, contrasted with annual operating expenses. The net value of capital assets is equal to the original purchase price or basis, minus accumulated depreciation.

Capital stock the stock issued and outstanding by a corporation representing investors' initial interest in the company.

Capitalization the means of financing an organization, consisting of equity capitalization (capital stock) and debt capitalization (liabilities including notes and bonds).

Cash flow the level of cash coming into an organization and leaving, used as a measure of management's ability to fund current operations.

Cash income a calculation of income without non-cash items (depreciation).

Cash-on-cash return a calculation of return in which net cash flow per year is divided by the initial cash investment.

Circumference of a circle the length of the outer edge of a circle, calculated by multiplying pi by the diameter.

Class lives the number of years over which assets can be depreciated, based on estimated useful lives of each asset class.

Comparative financial statement a financial statement in which two periods are reported together. These periods should be identical in length (month, quarter or year) in order to make a comparison meaningful; or should extend a current period to a year-to-date period.

Compounding method the calculation of compound interest. Methods include daily (using either 360 or 365 days); monthly; quarterly; semiannually; or annually.

Consolidated financial statement a financial statement including all segments and divisions of an organization.

Consumer Price Index (CPI) the most commonly used expression of the rate of inflation, published by the U.S. Bureau of Labor Statistics. The current rate of inflation is calculated by dividing the change in the index by the past index level (the index is based on price values in 1983 when the index value was identified as 100.

Contingent liability a liability that does not yet exist and may or may not exist in the future. For example, a company currently being sued may incur a contingent liability equal to the estimated damages it may have to pay in the future.

Conversion a change from one numerical expression to another. Conversion is needed to make functions easier between percentage, decimal and fractional forms of expression.

Cost of merchandise the cost factor used on the income statement. It consists of the beginning balance of inventory, plus merchandise purchased, minus the ending balance of inventory.

Cost of goods sold the dollar value of all costs, including merchandise, direct labor and other direct costs. The cost of goods sold is deducted from revenue to report gross profit.

Current assets all assets in the form of cash or convertible to cash within 12 months.

Current liabilities the value of all liabilities payable within 12 months, including 12 payments due on long-term liabilities.

Current ratio a calculation of working capital. Current assets are divided by current liabilities and the answer is expressed as a single numerical value.

Current yield on a bond a calculation in which the nominal (stated) yield is divided by the current value of the bond. Current value may be at a discount below or a premium above the bond's face value (par).

Daily compounding a method of compounding interest performed daily. The annual rate is divided by either 360 or 365 and accumulated on a daily basis.

Days' sales outstanding a ratio testing the trend in collections of receivables. The current balance of accounts receivable is divided by a full year's credit-based sales to find the average number of days' an account is outstanding.

Debt capitalization a form of capitalization in which the company borrows money and promises to repay the principal plus interest.

Debt coverage ratio a test of how effectively the current net income provides for payments of debts. The net income is divided by total annual payments to calculate the ratio.

Debt ratio a balance sheet ratio comparing debt capitalization to total capitalization (long-term liabilities plus equity capitalization). To calculate, divide long-term debt by total capitalization. The answer is expressed as a percentage.

Debt service the monthly payments required by the terms of a loan, including principal, interest and any additional payments required (for insurance, taxes, or loan processing).

Declining balance depreciation a method of depreciation in which the write-off during the earlier years is higher than in later years. The two most widely used methods are calculated at 150% and 200% of the straight-line depreciation rate.

Deferred credits items listed in the liability section of the balance sheet, representing future income not yet earned, to be reversed and booked as income in the applicable year.

Degrees of a circle the calculated degrees, used for preparation of a pie chart. The segments are first calculated on a percentage basis, and each is multiplied by 360 (degrees).

Denominator the lower portion of a fraction.

Depreciation a non-cash expense representing each year's write-off of a capital asset. The expense is offset by an increase to the accumulated depreciation account on the balance sheet.

Direct costs all costs directly attributed to revenue generation, including merchandise purchased, direct labor, and the freight cost to move goods.

Direct labor a form of direct cost for payments of wages and salaries, which are attributed directly to generating revenues and are expected to rise or fall in proportion to changes in the volume of revenues.

Dispersion (also called *spread*) in statistics, the degree of difference between the values in a field, and the average of the field.

Dividend yield the annual dividend to be paid, divided by the current price per share of stock, expressed as a percentage.

Earnings per share (EPS) the annual earnings reported by a corporation, divided by the total average number of common shares outstanding during the year.

EBITDA earnings before certain expenses; a calculation of net income adjusted by subtracting interest expense, taxes, depreciation, and amortization.

Equity capitalization shareholders' interest in the company, capitalization generated by investment rather than through debt capitalization.

Equivalents in multiplication logical statements drawing a logical conclusion. For example, when three values are considered (A, B, and C), if A is equal to B and A is equal to C then B must be equal to C.

Expenses non-cost and non-capital spending of a company during the year. Expenses are written off as necessary to the operation of the business. In comparison, costs are related directly to sales; and capital investments have to be written off through depreciation over several years.

Exponent (used in EMA) a calculated factor used in a weighted moving average, calculated by dividing 2 by the number of values in the average.

Exponential moving average (EMA) a weighted moving average that is easy to calculate and does not require adjusting the field each time to drop the oldest value and add the newest value.

Favorable variance in budgeting, the difference between the actual expense and the budgeted expense, when the actual is lower than the budgeted amount.

Field in statistics, a range of values. For example, a field of seven values is averaged by adding them together and dividing by the total of the field (7).

Financial section (report) in a report, a section or sections devoted primarily to numbers and dollar values. Examples include financial reports, productivity summaries, and budgets.

Financial statement a statement prepared to summarize current status (balance sheet) or the results of operations (income statement) and cash flow (statement of cash flows). The three statements are prepared as of the same date. The period covered by the income statement and statement of cash flows ends on the same date as the balance sheet's report of account balances.

Fixed asset turnover a ratio comparing annual sales to the average value of long-term (fixed) assets during the year. Sales are divided by the average and the results are expressed as the number of "turns" during the year.

Forecast a budget of revenues, in which assumptions are used to develop an estimate of revenues expected to be generated during the coming year.

Fraction conversion to decimal a calculation in which the numerator of a fraction is divided by its denominator to find the equivalent decimal expression of the same value.

Gross margin a percentage calculated by dividing gross profit by revenues.

Gross profit the dollar value calculated by subtracting the cost of goods sold from revenues.

Half-year convention in calculating depreciation, the first-year basis in which all assets within a specific class are depreciated as if they had all been purchased exactly half way through the year.

Home office expense the expense of operating an office in the home, including depreciation, utilities, insurance, taxes, and utilities. The calculation is based on proration of the full-time and exclusive use of the office space as a percentage of total floor space of the home.

Income statement a summary of activity during a specified period. It consists of revenue minus the cost of goods sold (resulting in gross profit), minus expenses (resulting in operating profit), plus or minus non-operating income or expenses (resulting in pre-tax net profit), minus provision for income taxes (resulting in after-tax net profit).

Income taxes a provision on the income statement located after the net pre-tax profit, used to calculate the "bottom line" or after-tax net profit.

Inflation the loss of purchasing power, expressed as a percentage change in an index from one period to another (Consumer Price Index).

Intangible assets those assets without physical value, including the assigned value of goodwill, covenants, and brands.

Interest coverage the calculation of the EBITDA divided by interest expense, as a means of identifying how well debt expenses are covered by income.

Inventory turnover an estimate of efficiency in inventory levels. It is calculated by dividing the cost of goods sold by the year's average inventory levels. The result is expressed as a number of turns during the year.

Liabilities debts and obligations of a company, divided between current (payable within 12 months) and long-term.

Liability-to-asset ratio a test of the relationship between assets and liabilities, calculated by dividing total liabilities by total assets.

Liquidity the level of cash available to fund current operations; working capital needed to fund expansion or to pay current expenses.

Loan amortization the process of repaying a loan, in which a monthly payment is calculated based on the loan amount, interest rate, compounding method and time to repay the debt; the calculation identifies the monthly payment needed to fully pay off the loan by the end of the period.

Long-term liabilities all liabilities payable beyond the next 12 months.

Lowest common denominator in a fraction, the lowest expression of the value. For example, ½ is the lowest common denominator for any fraction in which the numerator is exactly half of the denominator.

Margin of profit the percentage calculated by dividing net profit by revenue.

Mean in statistics, the average of a field of values.

Mean absolute deviation the absolute difference between an element and a given point. Typically, the point from which the deviation is measured is a measure of "central tendency," or the likelihood that outcomes will tend to be close to the mean or median of the field.

Median in statistics, the exact mid-point in a field of values. If there are an odd number of values, the median is the middle value. If an even number then it is the average of the two middle numbers in the sorted field.

Mid-month convention in depreciation, the method used for first-year depreciation. The assumption is that assets purchased during any given month were purchased exactly half way through that month.

Mid-quarter convention in depreciation, the method used for first-year depreciation. The assumption is that assets purchased during any given quarter were purchased exactly half way through that quarter.

Mode in statistics, the value that occurs most often in a field.

Monthly compounding a method of compounding in which the nominal annual rate is divided by 12, and the periodic rate is used to calculate monthly interest.

Moving average a calculation in which a set number of values are averaged in a changing trend. For example, a moving average of the most recent 10 months is calculated by adding values together and dividing by 10. The following month, the oldest value is dropped, and the newest value added to calculate a new moving average.

Narrative section (report) the report's section devoted to explanation rather than to financial or other numerical sections.

Negative cash flow the condition when more cash is being paid out than received, creating a drain on reserves.

Net after-tax profit the calculation of profits after all adjustments, including the provision for taxes; the "bottom line" of the income statement.

Net operating profit the profit before adjusting for non-operating income or expenses or provision for income taxes; gross profit minus operating expenses.

Net return the percentage of net profits divided by revenue; the best-known ratio based on the income statement.

Net return on equity a calculation of return to investors; net profit is divided by the net of the equity balance minus redeemable preferred stock, and the result is expressed as a percentage.

Nominal rate the stated rate of interest per year before calculating the periodic rate based on compounding method to be used.

Non-operating income or expense items added to or subtracted from profits that are not attributable to operations, including interest income or expenses, capital gains or losses, and currency exchange profits or losses.

Numerator in a fraction, the top half of the expression.

Payback ratio a ratio comparing investment to cash flow. The initial cash investment is divided by the annual net cash flow.

Percent of expense variance in budgeting, the calculated degree of variance. The dollar amount of the variance is divided by the budget amount, and the result is expressed as a percentage.

Percent of the total a common formula used in proration and other processes. The value in a field is divided by the total of all the values in the field, and the result expressed as a percentage.

Percentage change a calculated change in which an old new base is subtracted from a new base value, and the difference is divided by the old base value. The result is expressed as a percentage.

Percentage of revenue income statement component ÷ revenue

Periodic interest rate the actual annual rate with compounding included; for example, the periodic rate for monthly compounding involves dividing the stated rate by 12 and accumulating interest at one-twelfth each month.

Pi circumference of a circle ÷ diameter of a circle D = π 3.1416

Positive cash flow condition when receipts are greater than payments.

Prepaid and deferred assets classification of assets assigned to a later accounting period. A prepaid asset includes portions of the payment assigned to a later year; a deferred asset is not recognized until later as an expense or cost.

Present Value of a single deposit amount needed today to reach future amt.

Present value per period the amount needed today to fall to zero at the end of the period, based on the amount of each withdrawal, the number of months, interest rate and compounding method.

Pre-tax net profit $O +(-) N = P$

Price/earnings ratio (PE) Current price per share divided by the latest published earnings per share, resulting in a multiple (the number of years' net profits reflected in current price of the stock.

Projection an estimate of future revenues, costs or expenses, or of activity in cash flows.

Proof of proration prorated value of a + prorated value of b = known total

Proration division of a single value into two or more periods; for example, payment of an expense may be prorated between a seller and a buyer when a transaction is closed during the period applicable to the payment.

Quarterly compounding method of compounding over four periods per year, resulting in a higher annual interest rate than the stated rate; for example, 5% compounded quarterly results in annual interest of 5.0945%.

Quick assets ratio $(A - I) \div L = R$

Radius diameter $\div$ 2

Recovery period the number of years over which a capital asset is depreciated.

Remaining balance the dollar amount at the end of each year remaining payable on a debt.

Retained earnings on a balance sheet's net worth section, the accumulated equity from net profits or losses each year.

Return on cash invested (*Sales price – Purchase price*) $\div I = R$

Return on equity the investment return to shareholders or partners, based on net return as a percentage of net worth.

Return on net investment *(Sales price – Purchase price – Costs) ÷ cash invested = R*

Return on purchase price *(Sales price – Purchase price) ÷ P = R*

Revenue also called sales, the top line of the income statement; income before deducting costs or expenses.

Revised budget an annual budget adjusted based on variances, usually every six months.

Rule of 113 *113 ÷ interest rate (triple)*

Rule of 69 *(69 ÷ interest rate) + .35*

Rule of 72 *72 ÷ interest rate*

Semi-annual compounding a compounding method based on two periods per year.

Sigma the 18th letter of the Greek alphabet; lower-case Sigma, or σ is used in statistical formulas for the value of standard deviation.

Simple average an average without weighting or other adjustments, also called the mean.

Simple interest the calculated annual interest without compounding.

Sinking fund payment is a calculation of a series based on known assumptions. It answers the question, "How much money do I need to deposit per month, given a known rate of interest, compounding method, and time period, to accumulate a target amount in the future?"

Spread see *dispersion*

Square any number multiplied by itself and expressed as n^2.

Square root the inverse value of the square, the number that, when multiplied by itself, equals the principal value, denoted by the symbol $\sqrt{n}$.

Standard deviation the square root of variance, denoted by Greek letter *sigma*, or σ and denoted as $\sqrt{v}$.

Statistics a set of information based on numbers. As a process, statistics is the method of analysis. As a set of numerical values, statistics are the conclusions you get from manipulating those values.

Straight-line depreciation method of depreciating a capital asset calling for deduction of the same amount each year.

Tangible book value the book value (assets minus liabilities) of a company, without including intangible assets such as goodwill or brand value.

Time value of money amount borrowed, repayment term, interest rate, and compounding method

Total capitalization the combined funds to finance operations, consisting of shareholders' equity and long-term debt.

Unfavorable variance in a budget, an expense or direct cost higher than the amount budgeted or revenue lower than the amount budgeted.

Useful life the period of time in years, estimated for depreciation of a capital asset.

Variance (budget) $B - E = V$

Variance (statistics) square of each value in the field and then divide those results to find the average.

Volume of a cylinder radius^2 × π × *Height*

Volume of a rectangular solid $L \times W \times H$

Weighted average a method of averaging in which more weight is given to the latest entries in the field.

Working capital the net difference between current assets and current liabilities.

Year-to-date budget current-month budget + prior year-to-date budget

Year-to-date expense current-month expense + prior year-to-date expense

Yield measurement of profitability, alt. for *return*

Index

DOI 10.1515/9781547400638-016